Flexibility Measurement in Production Systems

Sven Rogalski

Flexibility Measurement in Production Systems

Handling Uncertainties in Industrial Production

 Springer

Sven Rogalski
FZI Forschungszentrum Informatik
Universität Karlsruhe
Haid-und-Neu-Str. 10-14
76131 Karslruhe
Germany
rogalski@fzi.de

ISBN 978-3-642-44146-2   ISBN 978-3-642-18117-7 (eBook)
DOI 10.1007/978-3-642-18117-7
Springer Heidelberg Dordrecht London New York

Cover design: WMXDesign GmbH, Heidelberg, Germany

Printed on acid-free paper

Springer is part of Springer Science+Business Media (www.springer.com)

Author's Note

In order to keep up in a constantly changing and evolving world, it is important to adapt to new demand swiftly. Dinosaurs dominated the earth for a long time until their sluggishness and inability to adapt to the changing climate led to their extinction. Other animals managed to sustain the various climate changes and followed the tides of evolution. Eventually, it was mankind that prevailed over all other life forms with its cognitive, affective, and social abilities, ruling the earth until today. The same basic principle applies to the business world where, in the long run, companies can only persist if they permanently evolve and react flexibly to changing environmental influences. For manufacturing firms especially, this means that they need to plan their resources in a way that permits them to stay cost and demand efficient effectively. In times of very high customer individualization and a vast variety of products and models, ever shorter product life cycles and frequent technological innovations, companies must find the right answers to the following crucial production related questions in order to persist on the global market:

- How can frequently changing customer requirements be analyzed and met?
- How can production systems be developed efficiently and reconditioned sustainably?
- How to make customized high-quality products within a tight budget and time frame?

Computer integrated manufacturing, virtual companies and integral production concepts are no longer enough to master the increased challenges of improving production flexibility, which are exacerbated by the current economic crisis. What is, in fact, needed is a holistic analysis of technical and organizational degrees of freedom in production. In this context, production managers are looking for new methods and tools to help plan, monitor and adapt their production systems, which include product life cycle and system related aspects in their aim for operating efficiency.

Sven Rogalski

Contents

List of Abbreviations and Acronyms

CAD	Computer Aided Design
Contol.	Controlling
CM	Contribution Margin
DCF	Discounted Cash Flow
ERP	Enterprise Resource Planning
MES	Manufacturing Execution System
MU	Money Unit
GUI	Graphical User Interface
JDK	Java Development Toolkit
KLR	Kosten- und Leistungsrechnung
OEM	Original Equipment Manufacturer
POC	Penalty of Change
PPC	Production Planning and Control
PSM	Production System Model
QU	Quantity Unit
SCM	Supply Chain Management
XML	Extensible Markup Language

Chapter 1
Introduction

"As the wind of change begins to blow, some build walls, others, windmills"

(Chinese proverb)

1.1 Importance of Flexibility in Production

Manufacturing firms have been facing the "wind of change", for several years, in the form of an increasingly complex and rapidly changing environment. The appropriate reaction to progressively dynamic markets and therefore different conditions for success are not "walls", presented by fixed company structures; it is, much rather, found in the activation of all available internal efficiencies and their use as flexible resources in competition, thereby seizing the resulting chances.

Production, no less than the place where resources and capabilities are transformed into products, has to make an active contribution to ensuring the company's long term prosperity [HaWh-88] [Schm-96] [AKN-06]. While product development, marketing strategies and financial power used to be considered the main factors determining competitiveness, production was attributed only a minor importance, until a reorientation process began in the early 1990s. This process was triggered by the Japanese industry's success with their very efficient use of production resources, which gained them continuously increasing advantages over competitors. This led to a new perception of production, and its image of an operative assistant shifted to that of a strategic actor, which is confirmed more than ever in today's age of globalization [ZaDi-94] [AKN-06].

The shifting importance of production was also followed by a change of production systems, which have been in a state of reorganization ever since [Chry-05] [AKN-06]. There were different approaches to increasing flexibility of production companies and facing the so-called adaptation-resistant factory structure [KRS-06]. Some traditional approaches are step-by-step product optimization, maximum planning of work processes in combination with division of labor, focus on the company's core competences, reduction of payroll related costs by investing in automation, and reduction of environmental pollution related costs. These basic approaches, however, assumed certain relatively stable environmental conditions, which are no longer commonly

S. Rogalski, *Flexibility Measurement in Production Systems*,
DOI 10.1007/978-3-642-18117-7_1, © Springer-Verlag Berlin Heidelberg 2011

found [Lutz-96] [LWW-00]; they have changed at an unprecedented rate in the last years. The globalization of the job markets and production sites, advanced by new developments in logistics and the internet, as well as the higher level of customer individualization, are said to impose the biggest of the new challenges. In this context, the term *turbulent environment of action* was established, which makes demand more unpredictable and therefore production harder to plan. Accordingly, all parameters of relevance for production, such as product structure, competitors, sales, and available technologies, can change rapidly and erratically. Thus, the predictability of changes in the industrial environment decreases on one hand, as the continuing product range extension and quick advance of technological developments like new materials, production techniques, or information and communication technology indicate [Warn-93] [LWW-00] [WHG-02] [EBGK-02] [AKN-06]. On the other hand, companies take longer to react to changes of their environment. This becomes apparent in the decision and work processes in production [WJR-92] [Rein-02]. According to Lindemann, the time from realizing the need for a change to its implementation nearly tripled between 1994 and 2005 [Lind-05].

Bleicher illustrates the resulting problem by defining the "Time Scissors" according to which, due to the increasing complexity, more and more time is needed to decide about changes and finalize their realization, while at the same time the greater dynamic of the companies' environment calls for quicker, more dynamic reaction (see Fig.1.1) [Blei-04].

As part of this constantly changing flow, companies are forced to find new ways of reacting to the various influencing factors precisely and in time. The corresponding cost and quality levels are just as important as the reaction time for maintaining competitiveness [RoKr-06a]. Therefore, finding answers to the core questions regarding adaptability of production systems, efficient handling of

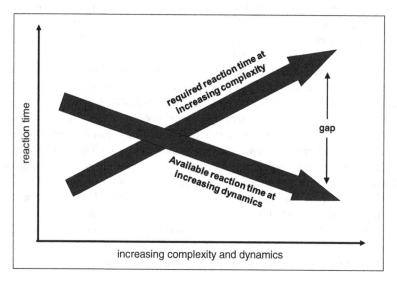

Fig. 1.1 Time scissors [Blei-04]

varying customer requirements and optimizing manufacturing processes is crucial [ARKO-07] [RSO-09] [Roga-10].

In light of this, reliable statements regarding a company's flexibility are of great value, as flexibility is an essential strategic factor. It represents production companies' capability of managing complex environmental situations, thereby increasing competitiveness and assuring long-term success [KaBl-05a] [RoKr-06b] [ARKO-07] [RoOv-09a]. Being able to gauge their production systems' flexibility allows production planners to evaluate their potential of adjusting to external influences, such as fluctuations in demand or product variations, as well as internal changes like additional capacities or staff assignment. Hence, invoking comprehensive flexibility evaluations in production management's planning and decision processes is a promising way of increasing the chance of production companies' successful existence in an ever changing competitive situation [RoKr-06b] [KRS-06] [RoOv-09b].

1.2 Current Trends in Handling Flexibility in the Field

Given the above-mentioned turbulent environment of action, the basic conditions for an efficient production and the importance of flexibility in this field have drastically changed. In this context, managing the steadily increasing planning uncertainty, regarding the type (range of products/product variations) and the amount (quantity) of manufactured products, has become a crucial competitive factor [KaBl-05b] [Niem-07] [ORK-07] [Roga-10]. As a reaction to this, production planners constantly strive to adapt their systems, strategies and concepts accordingly in order to gain sufficient agility to handle these uncertainties [KaBl-05b] [Bart-05] [KRS-06] [Roga-10]. A review of the last three decades will illustrate this.

While the market situation was still characterized by manufacturers in the 1980s, a trend towards more customer oriented production took hold in the early 1990s, triggering an ongoing sizeable expansion of the product palettes and variants [DoQu-04] [West-04] [Bart-05]. New company strategies emerged, which, after the diversification of production sectors, now focused on the core competences. Finally, production network management became the basic policy. At the same time, production strategies underwent a massive change. Computer Integrated Manufacturing (CIM) and Total Quality Management (TQM) were followed by the strategies of Business Reengineering and Lean Management, all still valid and in use, but enhanced over the years [Bart-05] [KRS-06]. Currently, Agile Manufacturing, mutability and Holistic Production Systems[1] are intensely pursued strategies, focusing on flexibility and other related qualities as important target values for adapting and improving production systems. They contribute to making production more flexible by including suitable concepts in the design and organization of production systems [Bart-05] [KaBl-05b].

---

[1]Holistic Production Systems bring together existing approaches to production strategies to form a new organizational model and can therefore be understood as methodological standards (operative guidelines) for the manufacture of products. They are not to be confused with technical systems for production, such as transfer lines or workplaces (vgl. [Bart-05]).

The currently dominating approaches to increasing flexibility in production, namely outsourcing, insourcing, acquisition of highly automated, reconfigurable production systems, increased inventory of resources and flex time, are now presented:

- *Outsourcing* of production can be seen as a useful and commonly found way to increase production flexibility. The key idea is the permanent transfer of whole production chains or parts of the production, both of which the outsourcing company used to perform itself, to external, independent companies. The substitution of in-plant production and the consequential supply of assemblies, modules and entire systems by partners allow the respective companies to keep their production more flexible and avoid high value-added costs by focusing on their actual core competences [Wild-05] [Bell-05].
- *Insourcing* is a powerful method for increasing production flexibility, used by several companies. Its idea is to spatially integrate important suppliers whose services are vital for production in the course of the company's concentration on its core competences. The responsibility for facilities and/or operational processes remains with the supplier, which relieves the insourcing company of the related tasks. This helps increase flexibility of production and reduce production cost at same time [Beye-04] [Bell-05] [Wild-05].
- The acquisition of *highly automated, reconfigurable production systems* is another way of raising the level of flexibility in production, practiced by some companies. Its main purpose is to increase productivity by providing a greater spectrum of possibilities and permitting shorter cycle times. A positive side effect of this type of flexibility boost is that in spite of or even because of the great initial investment and the related high fixed production cost (sometimes over 90%), an increase in labor cost hardly has any impact [Bell-05] [DHJM-06].
- The method of *increasing the resource inventory* is certainly among the most disputed when it comes to improving production flexibility, but still in frequent use. In order to be able to react to uncertainties of their environment flexibly, many companies accept high storage costs of raw materials and (semi-)finished goods as a sort of "vital reserve". This permits them to compensate fluctuations of demand and reduce costs and times of change orders. It is uncertain if this leads to capital commitment or results in a waste of potential changes of cost reduction [ZBM-06].
- In order to flexibly respond to temporal, intensity and capacity adjustments, companies often use different forms of *working hour flexibility*. Although this is not a new phenomenon, the different variations and potential freedom in the assignment of personnel allow for the fulfillment of different flexibility requirements. The aspects affected by this include the abolition of the until now customary separation of operating hours and working hours, the stronger connection between working and non-working hours or the increasing autonomy and self-regulation of the staff [KaBl-05a].

Practical experience has shown that, using these concepts, production can be organized more flexibly within a manageable scale [Wien-02] [KaBl-05a] [Bell-05] [WHK-06]. There are, however, still no sufficient instruments to determine flexibility deficits in production systems [KaBl-05a] [ZBM-06] [RoOv-09b] [RSO-09].

Flexibility-related characteristic values are currently not available or defined, which seems to be the crucial difficulty. Reasons for this can, for one, be found in the currently still unresolved problem of universally gauging and evaluating flexibility, which arises from the latter's multi-dimensional character [RoKr-06b] [DeTo-98] [KaBl-05a] [RSO-09]. On the other hand, flexibility requirements can vary in different parts of a production system, which calls for consistent, focused analytical methods. Such methods are currently not available [RoKr-06b] [GPMC-07] [RoOv-09a] [RoOv-09b]. In addition, the term "flexibility" is often associated with reconfigurable production systems in the field of manufacturing, and there are a wide number of scientific approaches based on their use in order to improve flexibility. This easily leads to the assumption that sufficient flexibility is gained simply by providing such systems, disregarding the context in which they are used.

For the above reasons, production system flexibility evaluations are rarely performed, despite their enormous practical relevance. At best, they refer to limited, industrial sector specific problems as part of prototypical research projects [KaBl-05a] [RSO-09]. At the same time, the steadily increasing planning uncertainty, combined with the current situation laden with heavy competition and high cost pressure, progressively forces companies to face the challenge of looking for new ways of holistic analysis of economic degrees of freedom in production. The key questions, therefore, pertain to the capability of reacting to fluctuations in demand and, more specifically, to what extent a change of the demand for particular product types or variants influences the operating efficiency of production as a whole. Furthermore, the possibilities of expanding the systems' capacities must be determined, because not only are changes in supply or demand of value from an operational viewpoint, but also of great strategic relevance [ZBM-06] [Niem-07] [RoOv-09a] [RSO-09]. Because of the lack of evaluation criteria, finding an optimal degree of flexibility that ensures competitive production is difficult and risky at the same time, especially before the background of the aforementioned concepts [SWF-05] [ZBM-06] [Roga-10]. If a company, on the other hand, out of fear of image loss or loss of sales establishes too much flexibility potential to assure that they could always guarantee supply, but ends up never using the potential, high and unnecessary additional costs that can eventually endanger profitable operation will be the consequence. The same goes for the opposite case, were insufficient flexibility potential causes repeated, uncoordinated and costly short-term system adaptations, which possibly results in the feared losses of image and sales.

The Fig. 1.2 illustrates the resulting problem of finding an economic balance between the predominating uncertainties and the appropriate degree of flexibility for production systems.

In conclusion, there is an ongoing, acute necessity for flexibility evaluation in the industrial field, with the interest of handling the following production related uncertainties:

- Volume – related fluctuations of demand
- Changes of the product/variant mix
- Capacity expansion requirements

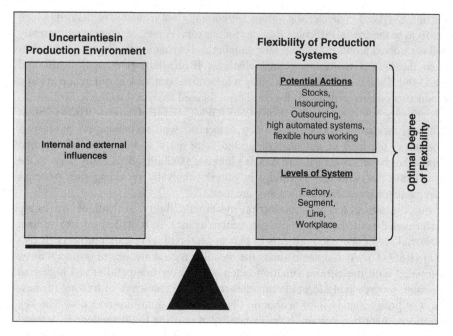

Fig. 1.2 Economic balance between existing uncertainties and flexibility of production systems according to Zäh, Bredow, Möller [ZBM-06]

For this task, industry-accepted evaluation methods are needed that do justice to the multi-dimensional character of flexibility, are based on common evaluation criteria and can be used across many industrial sectors.

1.3   Objective of This Book

Based on the given problem, a comprehensive flexibility evaluation concept is required that allows manufacturing companies to determine an economic balance between the predominating uncertainties and the appropriate degree of flexibility of their production systems. The aim of this book is therefore the introduction of innovative evaluation methodology that, determining characteristic values with interbranch applicability, yields economically reasonable conclusions about production systems' technical and organizational agility. The following types of flexibility need to be analyzed:

- Volume flexibility
- Mix flexibility and
- Expansion flexibility

In order to rate these three distinct types of flexibility, a variety of calculation methods are necessary, also called flexibility metrics. These have to accommodate the multi-dimensionality of flexibility and follow a common data and evaluation concept, thereby insuring the correct recognition and classification of dependencies between the flexibility types. It has to be possible to specifically attribute flexibility evaluations to chapters of production systems, because environmental turbulences have different impact on the different chapters. The sole focus of the analysis is on production with its direct and indirect processes; construction and development will not be considered.

In order to rate these three distinct types of flexibility, a variety of calculation methods are necessary, also called flexibility metrics. These have to accommodate the multi-dimensionality of flexibility and follow a common data and evaluation concept, thereby insuring the correct recognition and classification of dependencies between the flexibility types. It has to be possible to specifically attribute flexibility evaluations to chapters of production systems, because environmental turbulences have different impact on the different chapters. The sole focus of the analysis is on production with its direct and indirect processes; construction and development will not be considered.

In order to establish evaluation methodology as an efficient tool in production management, the user has to be able to quickly calculate characteristic flexibility values for specific system sectors. Also, the way in which the determined values are incorporated in the complete planned or existing system has to be easily understandable. A so called production system model is needed, which permits abstraction of relevant objects in production systems and their interlinking relations. This will make it possible to perform analyses of flexibility related dependencies between these objects and tracing flexibility deficits back to the responsible parts, thanks to the reciprocal links between the metrics and the production system model.

Finally, the suitability of the methodology as a system solution that can contribute substantially to increasing competitiveness of manufacturing companies in a turbulent market environment will be proven on the basis of a real implementation, which is also applied to a practical example.

The Fig. 1.3 illustrates the approach to tackling the given challenge.

The **guiding research question** in this book to solve the described challenge is as follows:

What criteria must a flexibility evaluation methodology which is accepted in practice follow and how is it possible to make the economically viable technical and organizational scope of production systems with regard to capacitive variations, Product/Variant mix changes and capacitive expansion requirements measurable?

Designed for production planners and managers in production, the book will provide reliable estimates of the flexibility of their production systems, which are vital in order to serve the market in time due to increasing complexity and dynamic as well as shorter reaction time. Thus, the evaluation methodology developed in this work is meant to be a supporting instrument for making decisions in operative and strategic production management that can be used to analyze and compare planned or existing production systems. In this context, the possibility to compare systems in different industrial branches is intended, such that analogies can be spotted and advancements transferred.

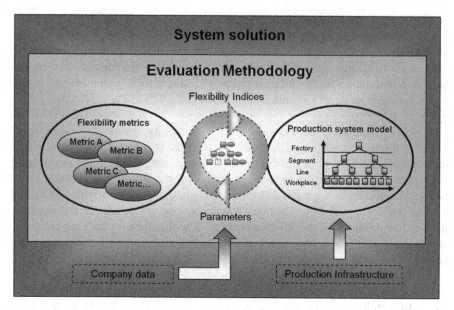

Fig. 1.3 Objective target

The *potential benefits* in case of a successful realisation of the methodology are:

- Unambiguous and transparent flexibility analyses through the use of consistent evaluation criteria
- Identification and containment of flexibility deficits in production systems regarding quantity-related demand fluctuations, changes of the product/variant mix and capacity expansion requirements by quantifying the scope of flexibility
- Support for short-, medium-and long-term security of an economical production by means of scenario considerations for the simultaneous increase in planning security and reduction of flexibility costs during market-related changes
- Reduction in the response time between recognising the need for production-related change and its implementation, due to flexibility analyses focused on various observation levels of a system
- Consideration of the multi-dimensionality of the calculation of production-relevant parameters, such as Break-even volumes or optimal production pro-grammes, to properly classify the identified flexibilities
- Identification of synergies between production systems in different sectors, due to a generalized flexibility calculation approach and the combination of flexibility metrics with a neutrally structured production model.

Chapter 2
Analysis of the Field of Observation

The aim of this chapter is to create a common understanding between the reader and the author and to determine how to handle the identified flexibility challenges through the proper containment of the object area. In accordance with the subject of this book, the first factors to be dealt with are the significance of the content, the influencing factors and objects, as well as the observation method used for production systems. In further steps, the flexibility of production systems in terms of their distinctive characteristics and meaningful classification options will be investigated. Here the relevant criteria for a goal-compliant flexibility assessment are highlighted. They are used for the accompanying evaluation of existing evaluation approaches and enable the establishment of the arising research demands. This all yields the necessary conditions for the requirements definition and design of the evaluation methodology.

2.1 Production Systems

The area of application for the evaluation methodology is production systems, for which there is a wide field of definition. To create a consistent and uniform denotation, the production is presented firstly as a system. In order for a deeper flexibility analysis to be performed later on, the characterisation of the system-describing resources and their flexibility-influencing properties is carried out. Because the resources can be organizationally summarised at various levels, the hierarchical structure of production systems will be discussed in a further step. This forms an important prerequisite for properly detecting the scope of the evaluation methodology. From this, a detailed examination is made of the factors influencing the production system and the resulting adjustment dependencies.

2.1.1 Definition of Terms

For a convenient derivation of the term "production system" it is recommended to consider the previous observation of the two terms "production" and "system", which are defined as follows:

S. Rogalski, *Flexibility Measurement in Production Systems*,
DOI 10.1007/978-3-642-18117-7_2, © Springer-Verlag Berlin Heidelberg 2011

Production can be generally understood as being the combination and transformation of production factors by certain techniques to form products. Production factors and products may represent both tangible goods and intangible goods (information, services, labour services) [Schm-96]. Based on the industrial point of view of production, the transformation of the primary and derived production factors available to the company takes place under the formation of combinations of factors in special production facilities (factories) in which large quantities of similar output per period are generated [Cors-99] [DCR-07]. This encompasses Westkämper's concept of production of pure production-related goods and services, along with all associated "controlling and organizational functions" that, according to Niemann, have an economic orientation [West-06] [Niem-07].

According to the system theory, a *system* is understood to be a sum of elements that display specific properties and relate to one another. It sets itself apart from its environment and is characterized by its qualities and capability to exchange material, information and energy with its surrounding systems. Systems organize and maintain themselves through their structures, where a structure describes the pattern/shape of the system elements and their networks of relationships through which the system works. A structureless compilation of several elements however, is called an aggregate. The system-related elements can in turn be viewed as independent systems, and the entire system itself, by expanding the observation horizon, can be a subsystem of a higher-level system (see Fig. 2.1). This results in a

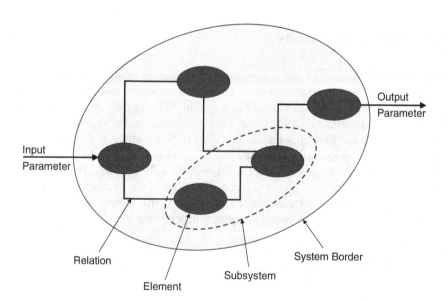

Fig. 2.1 General representation of a system according to Ropohl [Ropo-99]

hierarchical system view which makes it possible to choose an appropriate level of detail for a specific application [Ropo-99] [Trös-05].

In accordance with the underlying understanding of the terms production and system, a *production system* can be defined in the sense of Eversheim as an "independent allocation of potential and resource factors for production purposes", which in addition to the elements of the technical production process, also includes organizational elements for the planning and controlling of the production process [Ever-92]. It has cost accounting autonomy as well as an economical orientation [Ever-92] [Niem-07]. Accordingly, it has a specific system organization that creates specific links between the elements of a production system in order to achieve the optimal factor combinations to complete the task. The main influencing factors here are the number, direction, and the type or capabilities of the links [Kern-80].

2.1.2 Resources of a Production System

The resources as a group of adjustment objects in production systems are a key consideration in the planning, organization and control, and the associated competitive success of production systems. They are, in connection with the systems-view described in Sect. 2.1.1, to be considered as the system elements of which the overall system is composed. Following Penrose, a production system is to be understood as a resource bundle with the aim of achieving a profit. This bundle is fundamentally different in equipment, personnel and material [Gute-83] [Penr-95] [BeLu-03] [LES-06]. In accordance with their specific characteristics, the resources to fulfill the system's purpose are matched to one another as well as possible and define in their entirety the application range and the flexibility of a production system [Aggt-90]. Since the aim of the evaluation methodology is to quantify this flexibility, the three resources groups are presented in the following and are characterized by their flexibility-influencing properties.

2.1.2.1 Production Equipment

The resource "production equipment" is understood as being those basic factors of production meant to be used by humans to assist in the execution of tasks [PBS-05]. This includes all movable and immovable facilities and equipment that create the technical conditions for the generation of goods and services within a production system [MeBo-05]. According to the system theory they represent the technical elements. These include machines, tools, transportation and office facilities, as well as land and buildings [Gute-83] [MeBo-05]. In terms of their participation in the physical value creation/actual service delivery, they can be divided into *equipment for direct production involvement* (e.g. machinery, tools, jigs, etc.) and *equipment for indirect production involvement* (e.g. transport, premises and lands, storage facilities, etc.) [Aggt-90] [Zäpf-07]. Their common characteristic is that they are not consumed during their one-time use in the production process and can therefore be used multiple times [PBS-05].

Resources themselves have a special variety of uses known as their performance potential which influences the flexibility of a production system significantly. This potential can be described by the three parameters:

- Capacity
- Functional features
- Costs

Capacity is understood as being the quantitative performance ability of equipment as a producible output volume per unit of time, and there are also possible distinctions with regard to the maximum, minimum and optimum capacity. While the minimum and maximum capacities depend on technical and organizational quantities (e.g. the working hours of personnel), the optimal capacity identifies that output quantity where the implemented equipment used in the combination process is used most cost efficiently [PBS-05]. The operation, management and maintenance of the equipment depends on the staff, and it is through this that the application-/operation time of the equipment and thus the operating capacity is determined. While it is possible through automisation to achieve a decoupling, which opens up an additional flexibility potential, a complete resolution of this dependency can however not be achieved [Volb-81] [Zäpf-07].

The *functionality* of the equipment describes its technical scope of action. This refers both to the potentially executable editing functions (versatility) as well as to the processing quality (rating) of the services. Thus, for example multi-purpose machines usually have a high versatility for different goods and services production processes and sat the same time display a broad qualitative ability. In contrast, specialised machines are significantly reduced, since they often provide only one type of service, making their use versatility extremely limited, but the production result is often qualitatively superior [PBS-05].

The third relevant description quantity which indicates the performance potential of the equipment is the *cost*, which often depends on the capacity and functionality of the equipment. Therefore, for example, the purchase of an additional processing module for a machine leads on the one hand to an extension of its functionality, but also leads to additional costs. The capacity behaves similarly. Normally, the initial costs of two pieces of equipment (from the same manufacturer) are different if they, despite having the same functionality, allow different output rates. For this reason, the performance potential of the equipment is linked to its operating costs. The coverage of these costs is made in the form of scheduled depreciations, which are distributed proportionally over the years of use and intensity of use [Burd-02] [PBS-05] [Kale-04]. They have to be made on the basis of the original purchase price, including acquisition costs, where the relevant depreciation period of the assets is prescribed by the law (see [BMF-01]).

2.1.2.2 Personnel

The resource "personnel" means all paid employees belonging to a production system, which contribute to the fulfillment of tasks in some form. They represent the need for the human resources necessary for the service. Similarly to the equipment, the personnel are also distinguished by their involvement in the actual production in *personnel for direct physical production and services* (e.g. machine operators and rigging staff) and *personnel for indirect production and services* (such as accounting, inventory management etc.) [Aggt-90]. All personnel resources belonging to a production system have a certain performance potential, which in addition to the equipment also determines the system flexibility to a certain degree. The following three factors have an effect on this flexibility-determining potential:

- Qualification
- Working hours
- Costs

Qualification describes the cognitive, affective and physiological characteristics of a person which can be further developed through training and education [Lucz-93] [Penr-95]. It thus represents the functional spectrum of an employee i.e. his skills and abilities to perform potential tasks. The type of function as well as the performance of the person are relevant here [Volb-81]. The latter is directly related to the motivation, which can be influenced by appropriate measures, such as pay adjustments or task expansions [Lucz-93].

The *working hours*, the duration of which is determined based on the respective start and the subsequent end of working period, depends on various organizational constraints such as number of shifts and managerial or legal specifications [Brum-94]. The use of special working models allows a variation of working hours which provides the opportunity for increased flexibility of production systems, since they influence the total period of use of equipment. Shift work plays an especially central role in the flexible design of these operating times to achieve different working capacities [Schä-80] [MEK-05]. The following basic types are distinguished [CQP-89] [Bühn-04] [Hell-08]:

- *Non-continuous shift operation:* Characteristic of this is that the potentially available operating time of a working day (24 h) is not fully exploited and a work interruption occurs on the weekend. This applies to single shift operation, in which the operating time is limited to one shift (e.g. 8 h) per working day, as well as double-shift operation (e.g., two 8-h shifts per working day).
- *Partially continuous shift operation:* This is characterized by a 24-h operating time per working day with a break for the weekend. This is possible with a shift system of three successive shifts (early, late, night shift).
- *Continuous shift operation:* Here no organizational interruption of work is provided, which leads to a continuous operating time of 365 days a year (including weekends and holidays). This can be realized using four- or five shift operation, which often makes use of a so-called alternating shift model in which the shifts change in a certain pattern, e.g. weekly or monthly.

The third major determining factor for the personnel resource is the *costs* that have a decisive impact on the personnel planning within a production system [Volb-81] [Aggt-90]. They are directly related to the qualification, since an increasing level of qualification of the staff is associated with generally higher personnel costs. The same applies to working hours, the extension of which inevitably causes an increase in costs (e.g. due to overtime or increase in staff). Therefore, the flexibility of this resource is directly related to their costs. They result from the remuneration of the services provided by the staff's contribution in the form of a salary. This is divided into wages and salaries. *Wages* generally refers to payment by the hour (wage per hour worked) and depends on the hours worked in conjunction with the daily and weekly working hours. Conversely, *Salary* describes more consistent earnings (usually on a monthly basis) independent of varying working hours per month [Krop-01]. The trend shows that the stronger the relation of an employee to the actual physical production process, the more likely it is that he is paid for his work with wages as opposed to a salary.

2.1.2.3   Material

The third important resource in a production system is the "material". This term implies raw materials or goods that are, in combination with the previously described equipment and personnel resources within the production process, are processed to form finished goods.[1] Basically there are three different types of material. They fall into production materials, auxiliary materials and supplies [Kern-80]. The common factor of *production materials* is their ability to represent an essential component of the manufactured products. They include on the one hand, unprocessed natural resources directly obtained from nature, the so-called raw materials.

On the other hand it also includes goods resulting from pre-production, like materials (e.g. textiles and plastics, sheets, rods or pipes) and pre-fabricated components and assemblies, which are further processed into products of a higher order. In contrast, the term raw material is commonly used for *auxiliary materials* and *supplies*, since they are necessary for the execution of the production process. In this context, auxiliary materials are separated from the supplies in that although they are indeed part of the product, they are considered to be insignificant compared to the production materials. Examples of these would be adhesives, screws, rivets, electrodes or coatings. The supplies, in contrast to the other two types of material, do not go into the product. They do however ensure the essential maintenance and operational readiness of equipment. These are lubricants and abrasives, cleaning

---

[1]Final product is a manufactured item which has an existing a market for its sale (see Table 3.2, p. 57)

materials, coolant, office supplies and especially energy sources such as coal, oil, gas and electricity [Kern-80] [BBBD-03] [Dang-03].

There is a direct dependency between the consumption of the resource material and the number of products made in a production system. Therefore an increase in the production capacity creates new requirements for the provision of material, which demands a synchronization of the material supplier and the corresponding areas of production. The material is therefore highly dependent on the purchase order and this influences the flexibility of a production system through the following characteristics:

- Availabitlity
- Costs

The *availability*, which characterizes not only the time and place of the material use but also its type, quantity and quality, represents a substantial flexibility factor which significantly influences the production process. Thus an insufficient presence of material results in a restriction of the performance potential of the equipment and personnel and would consequently limit the scope of the production system. The responsibility here rests with the logistics, which guarantees both the internal and inter-company material availability and whose goal is to keep the inventory level in the stores as well as lead times[2] at optimal capacity utilization as low as possible. This usually causes a logistical complexity which should not be underestimated, because lead times, inventory and capacity utilization mutually influence each other [AIKT-04] [Zäpf-07].

As with the personnel and equipment, the costs also represent an essential flexibility-efficient criterion for the resource material. The reason is that the cost can have a significant impact on the availability of materials. For example, a high stock of material results in a high availability of this material, but also a higher capital commitment. Conversely, lower stock levels result in lower inventory costs, but they increase the uncertainty in the availability of materials, which can restrict the application range and thus limit the flexibility of a production system [AIKT-04] [Zäpf-07]. Therefore in the interest of a flexibly designed production process, the availability of the resource material must follow a cost-effective logistics concept.

2.1.3 Observation Levels in Production Systems

In the previous Sect. 2.1.2, the resources equipment, personnel and material, along with their flexibility-effective properties, were presented and explained. Their interaction and connections form the actual production system which fulfils the production task, whose link is the Organisation [Knof-91]. Through them each

---

[2]For more detailed definition of the term "Lead time", see Sect. 6.3.2

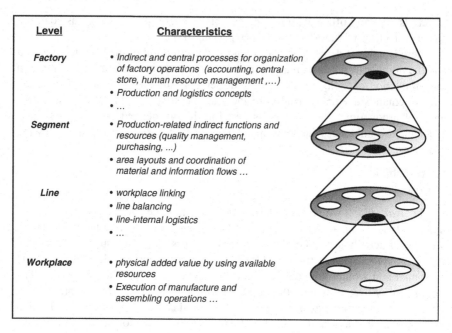

Level	Characteristics
Factory	• *Indirect and central processes for organization of factory operations (accounting, central store, human resource management ,...)* • *Production and logistics concepts* • *...*
Segment	• *Production-related indirect functions and resources (quality management, purchasing, ...)* • *area layouts and coordination of material and information flows ...*
Line	• *workplace linking* • *line balancing* • *line-internal logistics* • *...*
Workplace	• *physical added value by using available resources* • *Execution of manufacture and assembling operations ...*

Fig. 2.2 Observation levels of production systems according to Neuhausen [Neuh-01]

resource is summarized in functional system areas, which in turn can be constituents of other parent areas. This, is in accordance with the system theory, gives rise to a hierarchical breakdown based on Neuhausen [Neuh-01] which makes the classification in the levels *factory*, *segment*, *line*, and *workplace* possible. They are valid for this book. Sie soll für das vorliegende Buch Gültigkeit besitzen (see Fig. 2.2). Each system level stands for a specific set of system objects (for both the overall system and for its subsystems). Their assignment is based on fundamental, common, level-specific characteristics which are further described below.

2.1.3.1 Workplace Level

On the workplace level the actual service in the form of physical added value is performed under the combined resources of personnel, material and equipment and following a specific design principle. Through their number, their capabilities and their links with each other, the resources determine the operating range and the feasible processes of a workplace. We distinguish between manufacturing and/or assembly operations in the form of manual processing, machine processing with human assistance (semi-automatic) or fully automatic processing [DuOe-93] [Neuh-01] [BeHö-05]. The required lead time for manufacturing a product or commodity depends largely on the respective level of automation as well as the principle used in the machining processes. Normally a division into three basic

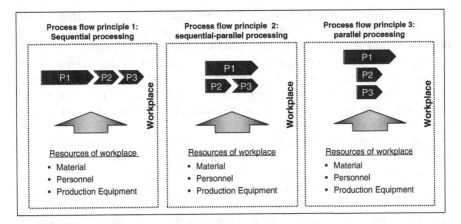

Fig. 2.3 Process flow principles for Manufacturing and Assembly operations on the workstation level

process flow principles can be made. Following they are explained in detail, while Figure 2.3 graphically summarizes these:

- The *Process flow principle 1* follows a workplace in which the processes run in a sequence, i.e. consecutively. Referring to the Fig. 2.3, a product would first have to go through the processes *P1* to *P3* before the processing of another can begin. An example of this could be a product processed at one processing station in the prescribed order of milling (*P1*), drilling (*P2*) and counter sinking (*P3*).

- The *Process flow principle 2* describes a workplace where parallel to the process *P1*, a product is simultaneously running through processes *P2* and *P3*. Following the example of the Process flow 1, at the start of the milling (*P1*) of a product, the simultaneous drilling (*P2*) of this product would also begin. On completion of drilling the counter sinking (*P3*) would take place, while the product is milled further. Such a process flow is considered to be sequential-parallel and can, compared to Principle 1, shorten the processing time for each product. The limits of this are the utilization rates of the process-dependant resources.

- In *Process flow principle 3* multiple processes at a workplace are running parallel in time. Compared with the other two principles, this is able to attain the shortest lead time, since according to the presented example, the three separate processing operations can be carried out simultaneously on the same product. In the field of fully automated processing this requires a precise coordination of resources and processing methods.

2.1.3.2 Line Level

The line level is characterised by the typical workplace linking for a line, in which the individual workplaces are linked together according to the material flow and information flow via so-called transport systems and transport means [Neum-96]. Since there is an equality or similarity of individual, sometimes even entire sections of the workflow sequence, the result is a strong product-based structure, in which the capacities of the often highly specialized workplaces of the manufacturing process are coordinated with each other [Aggt-90] [Schü-94] [Neum-96]. This combines different tasks, which ensure that the line is ready for operation and is functionally constructed, such as line balancing or line-internal logistics and controlling [Neum-96] [Neuh-01].

Differentiation possibilities of the line of principle result regarding the row- and the flow arrangement, which are distinguished from each other due to the time connection of the processing steps at each workplace, [Aggt-90] [Neum-96]. In the case of the *Row arrangement* there is no direct time synchronization between the individual workplaces, which is why it may sporadically result in waiting times or in a production with a workplace-related supply buffer. Accordingly, the transport system used must be suitable for a discontinuous material flow. A time specification exists only in a period-based theoretical volume performance. In contrast, the *Flow arrangement* concerns a work-cycle-bound row arrangement, whose time sequence is identified through the mechanised, continuous movement of products from one workplace to another. The workplaces and their connecting transportation systems are then spatially and temporally aligned [Ever-89] [Aggt-90] [Schn-01] [BCS-07].

Regardless of whether a row or a flow arrangement exists, lines follow a specified operation sequence of workplace chaining, which can be divided into three basic operation principles as shown in Fig. 2.4 [Phil-02] [Röhr-03]:

- *Operating Principle 1* describes the simplest organisational form of a line, whose technical elements (which included workplaces) are connected together in a directed non-redundant, sequential operation sequence. Thus, the line-related processing of a product begins with the workplace *WP1* and moves on to the next workplace only after completion of all manufacturing and assembly operations planned for the product at that workplace. This continues successively until the last line workplace. Thus, the capacity bottlenecks of a line are determined by those workplaces whose operations require the most time.

- *Operating principle 2* is characterized by a directed, redundant, sequential-parallel workplace linking. It is similar in essence to Operating principle 1, with the difference that workplace *WP2* as well as the machining processes performed there are laid out in double in the form of workplace *WP2a* and workplace *WP2b*. For example, should the productivity of a workplace belonging to a line following operating principle 1 be significantly reduced, this kind of workplace linking would make it possible to eliminate the capacitive bottleneck.

- *Operating principle 3* corresponds to a sequential-parallel arrangement of the workplaces, which enables both a redundant as well as a non-redundant control path. As shown in Fig. 2.4, a product variant *V1* could thereby be made on the upper operating path, while the manufacture of another variant *V2* as well as *V1* is carried out on a lower operating path. Thus the machining processes of the lower operating path are redundant when compared to the upper operating path,

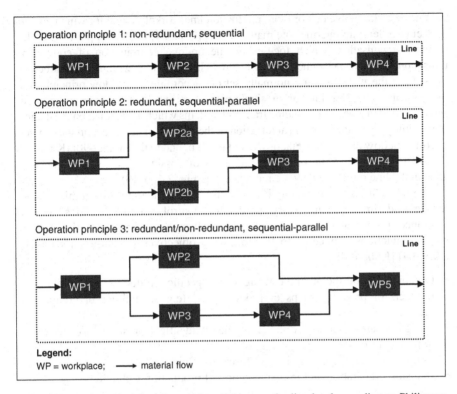

Fig. 2.4 Operation principles of workplace linking on the line level according to Philippson [Phil-02] und Röhrs [Röhr-03]

although this redundancy is not inversely valid. A product of the variant *V1* would therefore have the freedom to choose its control path, whereas a product of the variant *V2* would not. A possibility for such an operation principle would be a completely non-redundant arrangement of the parallel workplaces, when for example an additional product variant is to be manufactured.

2.1.3.3 Segment Level

The work areas contained within the lines can be assigned to the next higher level "Segment" and incorporated into a specific manufacturing or assembly structure. Segments, or manufacturing segments as they are also known, display self-sufficient, product-oriented organizational units/manufacturing areas made up of multiple levels of the value-creation chain. These units/areas of are formed according to specific product-market production combinations [Zahn-94] [Wild-98a] [BrGr-04]. The production resources implemented here are organizationally, spatially and structurally laid out in terms of a complete processing of parts in a value-creation chain. Segmentation criteria can be product types, sales structure, production processes or production volumes which involve one or more production line/workplace and for which an end market or at least an intermediate market exists [Lenz-04] [VaSi-04] [Zäpf-07]. The segment organisation, which focuses on the optimal coordination of equipment utilisation with the personnel, forms the link between and within the production units. Here the workflows are normally set out according to the flow principle, for example in a transfer line or in single-piece-flow production. Transportation is accomplished via the corresponding transport systems which, depending on the degree of versatility, specialization and automation of the segment, allow a continuous and/or intermittent flow of material with or without being bound to time [Aggt-90] [Knof-91] [VaSi-04] [Kobe-08]. The operation principles of possible linkages of production units in the line on which they are modelled (see Fig. 2.4, p. 19).

The relevant tasks concerning the level of the segment are: flow optimization to reduce lead times and inventory, production control, internal logistics, quality assurance, procurement and disposition as well as any other production-related indirect functions connected to the following objectives [Wegn-97] [Neuh-01] [VaSi-04] [Kobe-08]:

- Organization of the manufacturing of a specific production programme of products or product groups or product parts (e.g. assemblies or component groups)
- Design of area layouts by the grouping of identical or similar (in terms of process) products
- Coordination of the material and information flows
- Decision preparation and implementation of production technology

2.1.3.4   Factory Level

The factory level takes into account all technical and organisational elements of the production in a fixed environment, in which industrial products are manually and/or mechanically manufactured [Aggt-87] [Schm-95] [Spur-97]. This is done according to a specified organisational principle, which determines the relationships between and within the individual system elements and simultaneously arranges their classification criteria into various subsystems. The decisive criteria for this are the targeted business objectives of the overall system, subject to the system environment [Schm-95]. Regardless of the hierarchical structure of a production system, the factory level, as it is understood in this book, is to be regarded as the highest hierarchical level (see Fig. 2.2, p. 16) and thus defines the boundary of the system.

In contrast to the previous production levels, the indirect and the simultaneous central processes are prominent. It is through these that the organizational framework for the actual value creation is formed. They concern: accounting, human resource management, controlling, central store management etc. In addition the general construction is assigned major importance, through which the required linking of the processes with their spatial objects is achieved. It is closely linked to the production and logistics concepts of the factory [Aggt-87] [Neuh-01] [AIKT-04] [Kohl-07].

2.1.4 Change Drivers and Adaptation Objects of Production Systems

It is an unlikely probability that a production system may remain unchanged over its entire life cycle, from its commissioning in a specific configuration through to its decommissioning, and somehow still meet the diverse, ever-changing requirements; even if it is only designed for the production of a single product. The reason for the fluctuating requirements lies in the various external and internal influencing factors, or so-called change drivers, which shape the turbulent operating environment outlined in Sect. 1.1. The different types of change drivers are described in more detail below.

External change drivers are characterized by their common property of interacting with a production system from the outside. They may originate in technology, environment, resources or market changes and can be described as follows:

- *Technological drivers* are particularly new manufacturing technologies and procedures, materials, or IT technologies [WNKB-05] [Günt-05]. An example of this comes from the metalworking industry where the introduction of laser welding, due to its higher flexibility and efficiency, lead to the replacement of the conventional punch presses. The elimination of this punching process resulted in previously occupied work areas being freed up, storage capacities for the stamping tools falling away as well as the protective measures for noise becoming redundant. In return however, greater demands are placed on the ventilation of the new workplaces [WNKB-05].

- Changes in the *resources* personnel, equipment and materials can have an especially high impact on production systems [Dürr-00]. For example, the continued shortage of personnel would in turn lead to an increase in wages and thus would contribute to the concerned company seeking a higher degree of automation of their production systems.

- External change drivers allocated to the category *environment* are strongly influenced by political decisions and legal requirements, but also by social surroundings and the changes in the education and training systems [Wild-98b] [Dürr-00]. They result mainly in laws and collective agreements between employers and employees based on stringent safety and environmental regulations [WNKB-05].

- *Market-related changes* can be viewed as the most common change drivers affecting a production system. They can be traced back to the globalization of markets, the transformation from a manufacturer's market to a buyer's market and also the products themselves [Dürr-00] [RüSt-00] [Sest-03] [Bart-05]. This has, especially in recent years, led to a continuing increase in competition, which as a consequence forces production companies to bring new products to the market faster ("Time to Market"). This particularly illustrates the progressive expansion of the company's own, customised product range [Dürr-00] [Sest-03] [DoQu-04] [West-04]. The large increase in the number of variants, shorter delivery times and the large uncertainties regarding the required quantities and prices also account for the changing market conditions [Dürr-00] [KSW-02] [Rein-04].

In contrast to external drivers which have an impact on production systems, *internal change drivers* come from the companies themselves and are based on their objectives, strategies and performance deficits, as listed below:

- The *objectives* of a company can have a significant impact on the corresponding production system because they form the basis for its design and for the derivation of its technological objectives. Their influence is mainly aimed at those functions which lead to the system meeting its planned production tasks. A change in the company's goals therefore usually follows an adjustment of the production and thus the associated production system [Feld-91] [BiSc-95] [Dürr-00]. The resulting rearrangement affects not only the individual elements, such as the operational production functions, but also the overall system configuration. Possible reasons for the change of a company's objectives are either the target dynamics, mainly due to a different assessment of the goals by the decision makers or the uncertainty of the already defined goals, for example due to existing, unclear goal interdependencies [Kühn-89] [Dürr-00].
- In close relation to the company's intentions are its *strategies*, which imply the implementation of the objectives through appropriate measures. They are given in the company policy and determine the scope of the production system. Therefore, the strategies are considered to be major internal change drivers for the overall development of a production system [Feld-91] [Dürr-00]. For example, a differentiation strategy connects the high standard of the customer with the company-specific products, which have to distinguish themselves from other competitors through quality, innovation and individuality. Directly affected by this is the production system which has to meet the high requirements regarding the manufacturing and process technologies, the quality and accuracy of the processing or the nature and extent of the production functions [Frau-05].
- The *performance deficits* of a production system are also recognized as major internal change drivers and result from the system-specific vulnerabilities. These may emerge as a result of targeted investigations (e.g. internal analysis and benchmarking) or they may be identified in the Continuous Improvement Process (CIP). The analysis of hazards and the assessment of their potential threat in terms of company objectives then lead to corresponding changes in the production system. Examples of such vulnerabilities can be long lead times, low product quality, high production costs, low process reliability or confusing structures and material flows [Teum-04].

As a result of internal and external influencing factors, production systems are constantly subjected to pressure for change. To meet the resulting requirements, resource-based adjustments are inevitable. However, because the resources can be hierarchically summarized to specific system objects, which is in accordance with the concept of the system theory; the change drivers consequently affect the different

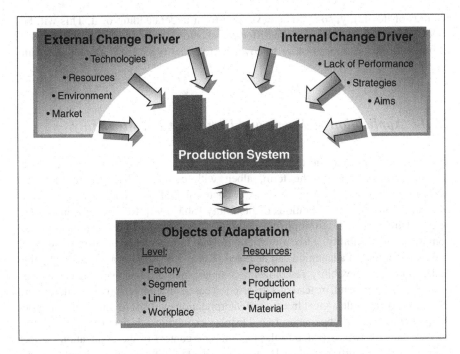

Fig. 2.5 Impact of internal and external change drivers on a production system

system levels, workplace, line, segment, and factory (see Sect. 2.1.3). An adjustment is therefore conceivable at all levels of observation in a production system.

Due to the different interactions between the different objects and/or different levels, system adjustments may not be made in isolation, but are always considered in context. This avoids sub-optimal and overall adverse system modifications, which may jeopardize the economic viability of production and the profitability of investments in production facilities. Flexibility measurements form the prerequisite for this, which on the one hand allow specific conclusions to be drawn on the need for adjustment and also the adaptation relationships of individual system objects organised in the various observation levels. It also identifies their impact on the system objects involved, including the overall system.

Figure 2.5 once again summarises the adjustment objects existing in a production system and the change drivers which affect them, before the flexibility as the subject of this work is further examined.

2.2 Flexibility of Production Systems

After having outlined the characteristics and the structure of production systems, in the following chapter we analyze the meaning of the term "flexibility" in connection with production systems. For this purpose, we first have to discuss the basic features of flexibility, so as to achieve a uniform understanding of it. This will be followed by an introduction to the different ways to classify flexibility (with regard to production systems), in order to cover its relevant types from a practical point of view.

2.2.1   Application of Terms and Dimensions of Flexibility

In vernacular language the term "flexibility" is used very often, which leads to a general understanding of the term, albeit with certain differences in meaning [Dürr-00]. Thus, in the Anglo-Saxon production-based technical literature alone, there are more than 70 definitions of flexibility [ShMo-98]. Reasons for this are to be found in the use of non-uniform terminologies, and an existing disagreement on the complexity of flexibility. Also, a restriction of the term to related concepts, such as agility and ability of adaptation can be noted [KeKe-05] [KaBl-05a]. That is why the understanding of flexibility resembles a complex multidimensional concept that is difficult to comprehend, instead of a description based on a universally valid definition [SeSe-90]. As a result, a great freedom of interpretation evolves: for one, it depends on the particular area of application (e.g. business economics, decision theory, or production) and for two, it is due to the fact that the context of the application can be interpreted in very different ways [Upto-97] KaBl-05a. With regard to the aforementioned, this book focuses on evaluating the flexibility of production systems. Thus, subsequently only selected definitions of flexibility are mentioned so as to provide a basic understanding of it, and to stand in line with the paper.

From a systematic point of view, to Pibernik flexibility is the ability of a dynamic, open, and socio-technical system to react in a goal-oriented way to relevant changes induced by the system or the environment, while at the same time it maintaining room to manoeuvre with regard to disposition [Pibe-01]. Schäfer describes it as a property of a system to adapt to changes of input parameters, basic conditions, or other influencing variables [Schä-80]. With regard to production, Horvárth and Mayer consider flexibility as the ability to improve the actual production on the short term, and keeping the operational possibilities open on the long term. At the same time, closely related domains such as production, sourcing, personnel management, finances and so on, make a contribution to flexibility, and thus also have to be regarded in balance with production [HoMa-86]. Similarly to that, Schmigalla also states the idea of room to manoeuvre, with defined options of operation and a fixed time frame. He further defines flexibility as the ability of a production system (considered to be constant in certain temporal limits) to adapt to changing demands of the production program and the technological process, without changing the number of its elements and its structure [Schm-95].

A common basic notion of these definitions – as well as of most definitions of flexibility – is the guarantee for room to manoeuvre, for decision-making, and for possible variations associated with the occurrence of change [Dorm-86]. However, it is not enough to just limit the flexibility of a production system to its ability to adapt, as the erratic and hard-to-predict demands of change that result from the insecurity of the environment can exceed the possibilities of a technical and organizational adaptation. Hence another basic aspect of flexibility is the ability to change, which enables a production system to adapt to changing conditions by altering its structure and the type and quantity of available resources [Wolf-90] [WHG-02] [KeKe-05]. KNOF considers this as the capability to act in order to achieve the anticipated goals of production, by which amongst others, chances in a sense of innovation can be created [Knop-91]. Consequently the flexibility of a production system is marked by its ability to adapt and change, both of which characteristics enable technical and organizational room to manoeuvre, in order to face insecurities of the environment.

This room to manoeuvre, which is available for covering the demands of flexibility, and which, according to Kaluza features a reactive and a proactive component, demonstrates an essential dimension of flexibility [Kalu-93] [Hall-99]. In the following, it is termed as the "Dimension of variety", and is one of a total three dimensions of flexibility that are required for its classification and description (see Fig. 2.6). In line with the characteristics of flexibility, the variety of a production system depends on the time and on the cost for its activation, which leads to the other two dimensions [Jaco-74] [Hall-99] [KeKe-05].

It is generally accepted, that flexibility causes cost. Hence, cost-effective charges have to be compared to an adequate benefit when building up potentials of flexibility [Hall-99] KaBl-05a. This causal connection clarifies the *cost dimension*, which encompasses two main aspects of cost: on the one side, the cost of keeping flexibility at disposal, and on the other, the cost of lack of flexibility. The cost of keeping flexibility at disposal is to be considered as a kind of *"insurance premium"*, which is to be applied only if need be [Wild-87] [Hall-99]. Examples are resources that are currently not in use (material, human resources and equipment), the cost of developing alternative plans, and also the higher cost per piece for multi-purpose machines.

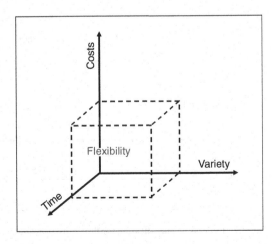

Fig. 2.6 Three dimensions for describing flexibility

As long as these flexibility potentials are not being made use of during the useful life of a production system, they do not leverage the cost of flexibility. That is why they are generally considered as idle time costs [Behr-85]. On the other hand, the cost of lack of flexibility that results from non-existing or non-sufficient flexibility potentials create a contrast. For one, lack of flexibility causes costs that stem from non-compliance of services – such as contract penalty, loss of follow-up orders and similarly problems. For two, they can cause additional cost, resulting from additional or missing adaptations of the production system [Meff-68] [Chry-96] [AMMC-05]. The latter occurs for instance, when there are no adaptation activities to suit a strongly decreasing demand of quantity on the long term. Such a limitation of production quantity can indeed be put into effect in every system; however it leads to a high cost per piece, which then jeopardizes the profitability of the production system. This example suffices to demonstrate the importance of the dimension of cost as one of the necessary descriptive factors of flexibility.

Another important parameter for characterizing the flexibility of production systems is the dimension of time. It plays a decisive role whenever the demand for flexibility is unsteady, because the building of flexibility potentials usually takes some time. In addition, potentials are to be considered as useless, if they are not available during a certain time period [Meff-68]. In this context we have to differentiate between short-term flexibility – which stands for the ability of the system to adapt –, and long-term flexibility, which aims at being able to change more fundamentally[REFA-90] [Hall-99] [Dürr-00]. Both short- and long-term flexibility have a close relation to delay time, which denotes the period of time between the occurrence of change, and the effective moment in which flexibility is measured (see Fig. 2.7). The delay time can be subdivided into times of cognition, awareness, decision, realization, and action [Bunz-85] [Behr-85] [Hopf-89]. From the point of view of the methodology of evaluation that is to be developed, one has to aim at the reduction of all five different sub-types of time; the extent of the reduction can vary in every single case or company.

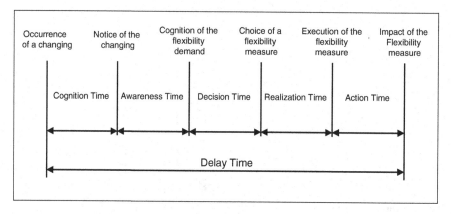

Fig. 2.7 Delay time classification for flexibility measurements according to Hopfmann [Hopf-89]

2.2.2   Classification of Flexibility

Besides the various notions of the basic aspects of flexibility as discussed before, there is also disagreement with regard to the classification of product system flexibility in technical literature. According to research by Sethi and Sethi, there are at least 50 different definitions for the various kinds of flexibility, several of which refer to the same kind of idea [SeSe-90]. Again, the reason for this is the use of differing terminology related to flexibility. Another reason is the lack of a standardized term and of a differentiated procedure, when flexibilities are being systematized and organized KaBl-05a.

A possibility of classification for different kinds of flexibility that is often found in literature, refers to system-related characteristics. For instance, Tempelmeier distinguishes eight subcategories of flexibility, which refer to particular system components, but also to the system itself, as is demonstrated in the following table (see Table 2.1).

Sethi and Sethi also classify flexibility in connection with system-related, adaptable properties. In total, they differentiate 11 kinds of flexibility, which they sub-divide in three groups: component-, system- and aggregate flexibility (see Table 2.2). Simultaneously, they correlate the different kinds of flexibility, so that their dependence on each other becomes more obvious.

There are great similarities between the classification system of Sethi and Sethi – whose classification and definitions of different kinds of flexibility are being quoted frequently –, and the principle of organization according to Koste and Malhorta. They differentiate between a total of ten systems of flexibility, which they also

Table 2.1 Different kinds of flexibility under the terms of Tempelmeier [Temp-93]

Kind of flexibility	Object of reference	Characteristic
Machine flexibility	Machine	Easiness of the adjustment to a new task
Material flow flexibility	Flexibility of material handling engineering	Transport of different products in any ways in the system
Working plan flexibility	Product	Possibility of alternative working plans
System modification flexibility	System	Modification of the number of resources
Flexibility of changings in the product mix	System	Possibility to change products without external modification of the system, but with setup activities
Product mix flexibility	System	Possibility to change products without setup activities
Cycle time flexibility	System	Possibility to manufacture products in different routes in the system
Flexibility of changings in the product quantity	System	Possibility to work economically at different flow capacities

Table 2.2 Different kinds of flexibility under the terms of Sethi and Sethi [SeSe-90]

Level of consideration	Kind of flexibility	Characteristic
Components/basic flexibilities	Machine flexibility	Different operation types that a machine can perform
	Material handling flexibility	Ability to move the products with a manufacturing facility
	Operation flexibility	Ability to produce a product in different ways
System flexibilities	Process flexibility	Set of products that the system can produce
	Routing flexibility	Different routes (through machines and workshops) that can be used to produce a product in the system
	Product flexibility	Ability to add new products in the system
	Volume flexibility	Ease to profitably increase or decrease the output of an existing system
	Expansion flexibility	Ability to build out the capacity of a system
Aggregated flexibility	Program flexibility	Ability to run a system automatically
	Production flexibility	Number of products a system currently can produce
	Market flexibility	Ability to the system to adapt to market demands

arrange into three domains of flexibility. There is the factory domain (quantity, mix, expansion, changing and new-product flexibility), a domain with regard to production (flexibility of operation and the operation chart), and the resource domain (human resources, engine, and material flexibility) [KoMa-99]. Compared to Sethi and Sethi, Koste and Malhorta associate flexibility primarily with an adaptation effort, which is economically efficient, and which causes little ramp-up or additional cost, and only an insignificant loss of the production system's output [SeSe-90]. Yet again, different kinds of flexibility are mentioned in different contexts, but once more they refer to the same idea; this happens often in related literature. Thus, Sethi and Sethi regard production flexibility as the same thing as Koste and Malhorta consider as new-product flexibility. Yet, both use a common terminology for machinery and operating chart flexibility.

Another method of systematization that is used by other authors does not classify different kinds of flexibility by its system-related adaptation parameters, but by a time-based view. For instance, Günther and Haller classify the time-related domain by operative and strategic flexibility. The same holds for REFA, only that this organization uses the expressions short-term and long-term flexibility, and that they do not describe the target value of flexibility, but its reference object instead, as compared to Günther and Haller [REFA-90] [GüHa-99]. The following table demonstrates the classification of different flexibility types of a production system, as proposed by REFA (see Table 2.3).

On the other hand, Kaluza uses his own approach to classification, by pointing out a basic difference between the objective of flexibility, (target-oriented) and resource flexibility. The target-intended flexibility refers to one that is considered to express new aims for the target system, the changing of this system (newly intended aims, for instance), and the modification of the level of target accomplishment as is regarded appropriate. On the other side, resource flexibility describes the availability of resources for achieving the aims that where set before. Furthermore, it is subdivided into real- and potential resource flexibility. The latter characterizes the ability of a production system to adapt in the domain of planning, decision, organization and control. In contrast, real resource flexibility – which also comprises quantitative and qualitative flexibility – characterizes the classic production factors on a physical level. At the same time, quantitative real resource flexibility stands for the ability of resources to adapt, and for physical work. Those resources are related to time, quantity and intensity. In contrast, qualitative real resource flexibility expresses the variety and the availability of retooling resources, and also a wide range of application areas for employees [Kalu-93].

The last type of classification – amongst the ones that are presented here – can be traced back to Wildemann. He makes a difference between quantitative-, qualitative- and time-based flexibility, and summarizes various characteristics of each kind, as is demonstrated in Table 2.4 [Wild-87].

Table 2.3 Different kinds of flexibility under the terms of REFA [REFA-90]

Name	Quantitative description	Timeframe
Product flexibility	Quantity of different work pieces; degree of freedom in regard to machine scheduling	Short term
Production redundancy	Quantity of operating material that can be applied alternatively	Short term
Volume flexibility	Economical limitation for addition shifts and short-term works; Keeping ready additional operating material	Short term
Adaptation flexibility	Effort of reconstruction	Long term
Expansion flexibility	Effort for retroactive expansions	Long term

Table 2.4 Different kinds of flexibility according to Wildemann [Wild-87]

Group	Quantitative flexibility	Qualitative flexibility	Time flexibility
Classification	Adaptation to changed quantities/structures	Adaptation to new production tasks	Time need for changing production tasks
Characteristic	• Ability to expand • Ability to compensate	• Variety, ability to retool • Production redundancy	• Cycle freedom • Automatic change over process
	• Ability to store	• Ability to reconstruct	

2.2.3 Relevant Criteria for Flexibility Evaluation on Production Systems

As emphasized in Sect. 1.2, the current challenges of flexibility and its application necessitate an assessment system which gives decision-makers the possibility to find a preferably optimal degree of flexibility, in order to identify the existing weaknesses. Criteria for assuring the useful evaluation of flexibility can be defined based on the conclusions concerning the production system that were drawn during the analysis and the linking flexibility. They can be explained as follows:

- Due to the hierarchic configuration of production systems, a focusable measurement of flexibility for different system properties is required, in order to detect characteristic variables in the domain of *factory, segment, line* and *workplace*.
- In order to meet the requirements of a multidimensional flexibility, the dimensions of *time, costs* and *variety* have to be considered equally when evaluation methods are being developed.
- To achieve a generally accepted assessment system for measuring flexibility, the system has to be *applicable across-industries*, and the *results have to be comparable*.
- In order to completely comprehend the assessment results, care must be taken to assess the flexibility in a way that is *independent from subjective influences*.
- As the crucial questions regarding flexibility of production systems in practice mainly refer to quantity fluctuations, changing of the product or variety mix and possibilities of capacity expansion; the domain of different kinds of flexibility that have to be interpreted, has to be configured in connection with the *quantity, mix* and *expansion flexibility*.

As demonstrated in the previous Sects. 2.2.1 and 2.2.2, the term "flexibility" is very broad and the inter-connected types of flexibility are based on an amplified and often diverse conception. That is why the types of flexibility that are considered to be important will be redefined in this book. This helps to establish a standardized understanding and when connected with the relevant criteria, it also forms the condition for the latter analysis of existing methods for flexibility measurements of production systems.

2.2.3.1 Definition of Volume Flexibility

The Volume flexibility describes the capability of a short-term and economical capacity adaptation for a given fabrication spectrum inside of the existing technical and organizational limitations of a production system.

Any flexibility metrics that follow this definition allow statements to be made about the scale of short-term, quantitative demand fluctuations that can be accommodated by the considered production system, without causing an economical deficit or demanding comprehensive extension of the system. Thus, the resulting level of flexibility is determined, based on an economically justifiable adaptation , by the characteristics of the actual existing elements and the structure of the system, as well as the system-linked organizational freedom of action.

2.2.3.2   Definition of Mix Flexibility

The Mix flexibility characterizes the stability and consequently the mobility of a production system in abandoning single products or variations concerning the production spectrum, without endangering the economical product fabrication.

This kind of flexibility assessment gives information on the harmonious composition of a production system with regard to production alternatives. At the same time, based on actual existing element quantities and on the system structure, an evaluation must be carried out on the extent to which the configuration of the product- and option mix may vary, without having negative consequences for the optimal production profit. In doing so, it can be shown which existing economical risk is linked to the system as a result of its dependence on particular products, options or elements necessary for the production.

2.2.3.3   Definition of Expansion Flexibility

The Expansion flexibility describes the ability of a production system to increase its capacity, by changing elements and/or the structure of the system in a targeted way while also considering the economical benefit.

According to the assessment of Expansion flexibility, the ability of a production system to enable adaptations to an ever increasing demand that exceeds the intended, short-term achievable capacity limits, has to be evaluated. Therefore, specific changes to the production system and/or elements (amount and characteristics) have to be conducted, while paying special attention to the expenditure, as it has to be compared to an adequate economical profit.

2.3   Analysis of Relevant Evaluation Methodologies for Measuring Flexibility of Production Systems

Comprehensive research in technical literature shows a great number of methods for the measurement of flexibility in production systems. This suggests on the one hand an active interest in this subject, but on the other hand it indicates the lack of a generally accepted operation method. Based on the criteria for a convenient flexibility evaluation that have been collected already (see Sect. 2.2.3), the large amount of evaluation methods could be reduced to a reasonable number. It is illustrated in the following preliminary work that is considered to be very applicable (further descriptions of operations can be found in Sect. 6.2) and analyzed in regard to these criteria. The aim is to draw conclusions about existing flexibility evaluation methods that might contribute to the solution for the design of an assessment method.

2.3.1 Evaluation Methodology by Zäh and Müller

Zäh and Müller suggest a design model that enables flexibility evaluation to assist capacities planning during insecure circumstances in regard to value creation in a factory. Its purpose is to help decision-maker to judge available production capacities in regard to their qualification of handling expected market fluctuations and to its profitability, as well as finding the right strategy for a comprehensive capacity optimization of the production system. The application of the design model takes place in three main steps whose first starts with a prediction calculation. Here, based on past data a demand prediction occurs, that comprehends the trends, seasonal dependency, as well as potential distortion, and thus, judges the influence of possible market fluctuations. During the second step for each product/each production option the accordant and company intern value creation process is modeled, whereas each process is described by its relevant, process specific parameter (e.g. processing time per piece, set-up time per operation, number of shifts), plus by its so-called global-parameters (e.g. labor time per shift, number of variants). In this way, process related flexibilities can be pointed out with reference to its capacity, and in dependence on different production- and amount configurations. The third and last step comprehends the setting up of a cost structure that helps to identify a variation specific profit, by applying the calculation of Contribution margin. Through comprehension of the calculated prospects values, an expected total profit can be calculated, that gives information about the optimization potential of the analyzed production system, and has to be considered as a substitute for its flexibility [ZäMü-07].

The so-called "capacity-/costs-design model", suggested by Zäh and Müller, considers different products/product alternatives, processes and their configuration, as well as dynamic supply shortfalls. The actual basis for evaluation is the economically optimal production capacity, which indeed enables estimation on expansion- and Mix flexibility, however, not on the base of concrete calculation. In fact they refer to Volume flexibility that is judged on the basis of the discrepancy between the actual arrangement of the production system with its functions and demands that are expected in the future. At the same time the simultaneous consideration of the three dimension costs, time and variety is specially emphasized. Moreover the operation is applicable cross-industrially, mostly independent from subjective judgment of the user and allows comparison between different systems. However, as it looks about the company internal value creation, in the first instance it is suitable for judgments on a factory level. Flexibility statements about segment-, line- or workplace level can only be deduced in an indirect way in the context of the analysis.

2.3.2 Evaluation Methodology PLANTCALC™

On the context of a joint research project between the TU Munich and the Siemens AG a method for evaluating flexibility in production was developed, that has been realized technically with a software called PLANTCALC™ already. For a production system that has to be analyzed the predominant flexibilities have to be determined, as for example mixed-, process- or Volume flexibility. Based on an existing scale they are being evaluated in regard to their uncertainty and their meaning as an answer to fluctuating environmental influences, by what flexibility deficits can be identified. In the following the step an Event Tree is modeled, that is put together by different environmental conditions, which are expected with different likeliness, and, according to this, represents the processing of production respecting demand, product mix, prices etc., as well as the therewith connected requirement of flexibility. In dependence on the relative environmental conditions thereupon the most economical operating mode of the production, based on production costs and the costs for modification in another mode, can be calculated. The technical realization of this evaluation method takes place with the simulation software PLANTCALC™, which, considered as a flexible expandable platform, has to enable the cross-industry application of this method [ZBM-06] [KoKr-08].

This procedure was engineered following to the research project and realized in PLANTCALC™, and thus, is suited very well for flexibility orientated evaluations in the context of factory planning, that has the purpose to make flexibility statements on future production developments based on costs and stochastic functions. It can be applied cross-industrially and allows comparison between different systems. However these kinds of consideration take rather place on factory level, while the level of segment, line and workplace are in fact considered, but are not named explicitly. Similarly applies for the considered flexibility kinds quantity, production mix and expansion. Although they are a part of flexibility evaluation, no clear conclusion can be drawn about them. Furthermore, time is not considered in a sufficient way, as the procedure, due to its planning character, comes from a long-term time frame. The subjectiveness of the evaluation results has to be considered critically, too, as the emphasis of the involved kinds of flexibility depend on the evaluation of the user, that is reproduced on a one-dimensional scale.

2.3.3   Evaluation Methodology by Wahab, Wu und Lee

Wahab, Wu and Lee from the University of Canada have another approach for measuring flexibility cross-industrially in production. Here they combine the efficiency-orientated flexibility of a machine with its time-and cost related flexibility with a particularly developed framework. In a first step time- and cost charges of a machine are being evaluated with the so-called "DEA (Data Envelopment Analysis)-Model" in regard to its possible job realization and the correlated configuration (time-/cost related flexibility). In the following second step the output based machine flexibility is included in the calculation, too, by adding technological aspects like the number of possible operation procedures, their relevance and the probability of their procedure. Based on this input-information, and using the "flexibility"-model that is integrated in the framework, the final evaluation of the machine flexibility is carried out, based on time, costs and variety [WWL-05].

The framework presented by Wahab, Wu and Lee is in principle appropriate for the calculation of the quantity-, and variety flexibility on a machinery level. One of its strength is demonstrated in the simultaneous consideration of costs, time and variety, as well as the cross-industry application. Furthermore, comparison with different systems can be executed, with the exclusion of subjective influencing factors. However, a great disadvantage is the exclusive focusing on flexibility evaluation of machines. With the existing conception of this framework, an application on line-, segment-, or even factory level is not possible. In addition to that, possibilities for considering Expansion flexibility in a production system are missing.

2.3.4   Evaluation Methodology by Schuh, Gulden, Wernhöner and Kampker

Schuh, Gulden, Wernhöner and Kampker present a concept for the setup of a system of indices that has to enable flexibility evaluation beginning from the workplace- via line-, segment- and factory level and to the point of production network. Referring to the object concentration that is known from software-engineering, the setup of a category based system model in connection with the production system that has to be considered is possible, by describing each actual existing object (workplace, line etc.) by a class. Such a class contains parameters for their description such as kind and number of products or set-up- and processing time, as well as operations for flexibility calculation on the particular organizational level. With the help of the principle of inheritance, a planner is able to determine individual assessment items relatively fast, as for example a line, and can link it with the system of indices. The indices system itself consists of three parameters that represent quantity-, variant- and product changing flexibility.

Their calculation occurs with specially designed, flexibility dependent algorithms with the consideration of time, costs and variants. The calculation rules, that base on these information refer to particular class parameters of the system model, as well as to necessary reference scenarios, that have to be defined by the user preliminary [SGWK-04].

The system of indices, designed by Schuh, Gulden, Wernhöner and Kampker is mainly characterized by its dynamic setup, which leads back to the object-orientated system concept. Thus, application specific configurations on the system of indices can be established pretty easily, in order to conduct differentiated flexibility evaluations. These evaluations should occur on different organizational levels of production systems, and independently from various sectors. However, it should not be disregarded, that the applicability of the system of indices was confirmed on a workplace- and line level only. The possibility to regard flexibility on a segment-, factory or even production network level is indeed possible on a theoretically level; however, the actual realization remains unsettled. Furthermore it has been discovered that both quantity- and Mix flexibility are assessable. Expansion flexibility has not been considered, however it could be deducted from the represented production changing flexibility. All the same the main point of criticism is the dependence of the evaluation procedure on subjective factors. Thus the user of the indices system has to appoint scenarios concerning the market development that he has to judge in regard to their importance and occurrence probability, and also in accordance to his experiences.

2.3.5 Evaluation Methodology by Ali and Seifoddini

The procedure for evaluating flexibility of production systems represented by ALI and Seifoddini is basing on a combination of product mix and production capacity. The items that are being considered are engines that are available for fabrication processes, manpower and logistics. Here, their main influencing factors like quality, time, availability, knowledge, age etc. are being documented with imprecise parameters, such as "High", "Medium", or "Old", "New", "Very old" and so on. By applying concrete fuzzy-logic-rules and simulation environment, reactions of the production system to demand fluctuations, to changing of production capacity and product mix, as well as to changing human resources demands, working shifts and diversified workplace- or personnel capacity, can be assessed using simulated test runs [AlSe-06].

With their fuzzy-rules-based procedure, Ali and Seifoddini enable considerations with regard to mix- and Volume flexibility on an engine- and line level. They are transferable to a segment- or factory level by the use of appropriate flaring. However, the use of this procedure depends on the particular industry and concrete application. Furthermore, evaluations on Expansion flexibility cannot be executed. Another disadvantage is the negligence of cost considerations, that prohibits economical considerations of different production systems and accordingly of different production system configurations. In addition to that the time dimension is regarded in an insufficient way due to simply qualitative considerations ("High", "Median", "Low" and so on). Another deficit results from the subjective consideration of relevant influencing parameters with imprecise characteristics, which makes the significance of the operation and the comparability of its conclusions depend on the user's appraisal.

2.3.6 Evaluation Methodology Desyma

DESYMA (Design of Systems for Manufacture) is the name of an flexibility evaluation method that has been developed at the Greek University of Patras. It approves the evaluation of alternative design solutions of production systems under the consideration of demand insecurities. Here, flexibility is being identified with statistical analysis of discounted estimations of the Cash-Flow, the so-called "Discounted Cash-Flow" (DCF). In this case of flexibility, the life time cycle of a considered production system is evaluated for a certain time frame and different market scenarios. With the underlying algorithm, the consideration of pioneering costs of the production system that has to be analyzed, as well as the consideration of situation dependent adaptation costs and the resulting variable production costs, occurs per period. Based on the distribution of resulting, situation dependent DCF-values, the evaluation of flexibility within the predefined market environment occur afterwards. There is a correlation between DCF-values and the production system: the smaller the distribution of values, the more flexible is the production system, as it is less prone to assumed market scenarios, and thus, life cycle costs are more stable. [ABMC-05] [AMPC-07].

The advantage of the DESYMA-method is its cost-related evaluation of production system related configuration alternatives. Due to Cash-Flow-analyzes for different market scenarios and involved pioneering costs, particular costs (for adaptation measurements, for instance) can be evaluated cross-industrially with the corresponding profit (by sales exposure). With the help of the specific dispersion of DCF-values, a kind of "collective flexibility" is being described for the production system that has to be assayed. The scenario related part of this method turned out to be disadvantageous. It is the reason why the quality of the results depends on the subjective user's judgment. This procedure focuses mainly on the workplace- and line-level, whereas the segment- and factory level can be involved in parts, too. Furthermore it has been established, that with the help of the specific dispersion of DCF-values, the Expansion flexibility of a production system can be deduced in fact pretty easily, however, no direct analyzes in regard to quantity- and Mix flexibility are possible, although scenario based quantity- and product mix fluctuations are included in the DESYMA-method. Another deficit is the emphasis on a planning orientated procedure, which leads to the fact, that exclusively long-term flexibility evaluation is possible.

2.3.7 Evaluation Methodology by Peláez-Ibarrondo und Ruiz-Mercader

Peláez-Ibarrondo and Ruiz-Mercander execute an evaluation of flexibility with the help of an integrated calculation of quantity- and Mix flexibility, which they call "operational flexibility". Assuming that production systems are available discretionarily and that the production spectrum for Flow-Shop-production-systems that has to be investigated is scenario-depended, flexibility is evaluated on the basis of the difference between the minimal acceptable and the maximal possible production rate. The former is determined by the Break-Even-Point,[3] while the maximal production rate is limited by the engine with the lowest maximum capacity in the assayed production system, which consequently represents its supply shortfall. In order to determine required parameters, the evaluation method includes production relevant fixed- and variable costs, as well as essential time aspects, like production- or working-time. The actual evaluation of the work plan-flexibility for a considered production system results from a scenario caused investigation of quantitative configuration-variations of the fabrication spectrum within a defined acquisition period [PeRu-01].

The procedure of Peláez-Ibarrondo and Ruiz-Mercader basically seems to be appropriate for evaluating quantity- and Mix flexibility, even if it they are expressed only in an indirect way in regard to the current application. However, possibilities for evaluating expansion-flexibility are completely missing. A great advantage is the consideration of the Break-Even-Point, that does not only involve fixed-, but also production dependent variable-costs. However, one-product-engines, that produces according to the "Flow-Shop"- model, are being considered only. Thus, solely the same engines in the same order carry out the required work-operations for production, which limits the sector-independence of this procedure. Besides, it remains open, how flexibility for different work-operations at different engines has to be reviewed, especially whenever several-product-engines can carry out intersecting operations. Furthermore, previous flexibility considerations simply occurred on the workplace- and line-level. The applicability for the levels segment and factory is questioned due to its insufficient consideration. Another criticism concerns the lack of precise index numbers in regard to wok plan-flexibility, which enable comparison amongst different system. Instead of that, Ibarrondo and Ruiz-Mercader only name the quantity dependent fluctuation rates and the particular short-fall for each scenario in form of absolute numbers.

2.3.8 Evaluation Methodology Penalty of Change

Due to the property of flexibility of a production system, to act opposed to its sensitivity in regard to changing, Chryssolouris operates a kind of flexibility

---

[3]The Break-even-point describes that sales volume that covers all costs with the sales revenue (break-even), so that the profit is zero [Eber-04].

measurement, that is based on the evaluation of additional costs (penalty costs) in case of an insufficient flexibility. For this he uses the "Penalty of Change" (POC) as an index. In case that considered changing within a system can be executed without any further costs, the system holds maximal flexibility, so that the POC-value is zero. If, in contrast to that, cost causing changing are necessary, the system is considered to be more inflexible which causes a higher POC-value with increasing costs. For calculating the POC-value, Chryssolouris goes back to the concept of the Event Tree, with what he calculates particular costs on the basis of occurrence probabilities of future scenarios. The scenario that totally features the lowest cost is considered to be the best [AMMC-05] [GPMC-07]. Consequently a one-product-engine that is able to produce 1,000 pieces of product *A* per day and that causes costs of €20,000 with a realization probability of 80% has to be considered more flexible than an engine, that features a realization probability of 20% and costs of €100,000, but that is only able to produce 10,000 pieces of product *A* per day. The so-called penalty costs make a difference of €4,000 Euro in benefit of the cheaper engine.

The application of flexibility evaluations on the basis of the "Penalty of Change" is possible from an organizational point of view that includes all processes from the workplace to the point of the factory, and specially makes sense for the evaluation of different options or expansion alternatives in regard to costs. A crucial disadvantage is the distinct dependence on occurrence probabilities of scenarios and the corresponding costs. Thus, the quality of the results is dependent on each user's objective ability of judge, which can lead to contrary choices. If the user presumes that the occurrence probability of both scenarios as shown in the previous example of the two one-product-engines is 5% higher for instance, he will choose the more expensive engine, as this scenario causes lower penalty costs (€2,000 less). Another criticism results from the disregard of time and the lack of evaluation of actual mix- and quantity-flexibility in a system. They only can be provided in connection with regarding the future, whereas they are limited to qualitative information, in terms of: "Engine *A* is more flexible as engine *B*, due to higher penalty costs of *A*". Furthermore, this procedure does not offer reasonable comparison between systems of different dimension although it can be applied cross-industrially, as the POC-value is quoted in an absolute number (relative penalty costs).

2.3.9 Evaluation Methodology by Haller

Haller provides another procedure by evaluating flexibility of automated flow of material systems that are part of varying high volume productions. He differentiates between aims of the operational and strategical flexibility, whose importance is assessed in dependence on their usage. This occurs with the help of criteria for describing system properties (e.g. processing- and shutdown time, or the influence of the product mix on the flow capacity) and system requirements (e.g. expansion of the delivery rate or dependence of secondary resources), which describes the stability and robustness (operational flexibility), as well as the ability of integration and adaptation (strategical flexibility) of a system in regard to changing of quantity, variants or products. In order to maintain the totality and the objectiveness of the evaluation, certain indicators are being used in order to ensure a high transparency for evaluating the criteria. At the same time HALLER also considers quantitative criteria in his model by using indices of the profitability. Here monetary effects of flexibility are expressed on the basis of dynamic investment calculations. Based on an adequate summary of single results of qualitative and quantitative evaluations, appropriate conclusions concerning flexibility can be carried out afterwards [Hall-99].

With his dual evaluation model that includes the analysis of the utility value and costs/investment calculation, Haller offers a proper base for researching flexibility of different production systems, even if it is specially orientated on the flow of material. Due to a configuration that is universally valid, this model should be applicable in other branches or domains of production if negligible modifications are carried out. Indeed qualitative evaluation criterion play a decisive role in this procedure, however using the provided indicators, an accordant degree of objectiveness is ensured. Furthermore, the possibility of evaluating all organizational levels, from workplace to factory, has to be pointed out. However, the involvement of time is a weak point, as in opposite to the dimensions costs and variety it is considered in an insufficient way. To all intents and purposes time aspects are part of the evaluation, like processing- or holding time, however their importance is reduced to qualitative information and to a part of the investment calculation, as for example for measuring the interest rate or the calculatory return flow. There is no concrete reference to time in order to calculate a range of flexibility. Another great deficit of Haller's procedure is the lack of an explicit model with that quantity- mix and Expansion flexibility can be analyzed, whose cause is the strong focus on alternative evaluations of production systems.

2.3.10 Evaluation Methodology FLEXIMAC

Another way of evaluating flexibility is the FLEXIMAC method. It goes back to the analogy of dynamic properties between a mechanical system and a production system. According to that, the ability to react on needed changing in a mechanical system is characterized by a so-called *damping factor,* that can be calculated under certain circumstances using a transfer function [APMG-06]. The same applies for a production system whose flexibility, according to the developer of the FLEXIMAC-method, has to be evaluated with the ability of reaction in regard to the dynamic input. It is presumed that input is the intended processing time of a system that is necessary to produce a varying number and kind of products, whereas the output represents the actual flow time. In the ideal case, input and output are identical, as no delay, as for example because of rig procedures or lacking resources for the factory, occur. Before the FLRXIMA-value can be calculated it is necessary to divide a given period in equal time intervals, and the processing time as well as the flow time have to be opposed. After that, the processing time and the flow time have to be added up so that a discrete Fourier transform can be carried out and a transfer function can be determined by calculation of quotients. Based on this and according to the damping factor of mechanical systems, the FLEXOMAC-method calculates the eigen-values of the system and the amplitude with the frequency of the eigen-values. The concrete FLEXIMAC-value results from the arithmetic mean of the ten highest amplitudes, as mentioned in the following formula. The higher this value for a production system is, the more flexible it is, as it is less sensitive towards changing of the input [APMG-06] [GPMC-07].

$$\text{FLEXIMAC} = \frac{1}{10} \sum_{i=1}^{10} \left(\frac{1}{2Q_i} \right)$$

The great advantage of the FLEXIMAC-method is its independence from subjective assessments and reference scenarios. Furthermore, flexibility evaluations can be executed cross-industrially on all relevant levels, starting at the workplace to the point of the factory. Data that is relevant for the calculation can be gathered from working data or Enterprise resource planning (ERP) Systems pretty easily. A remarkable disadvantage is the disregard of costs, although the dimensions of time and variety are completely regarded. In addition to that the FLEXIMAC-value only allows a general valuation of the analyzed flexibility. Concrete conclusions regarding quantity- or Mix flexibility are possible in a limited way only, whereas an appraisal of Expansion flexibility is not at all possible. Another deficit of this method, that should not be underestimated, is the orientation on the past. This requires the actual existence of production data of an analyzed system, so that objective flexibility measurements can be carried out. That is why comparison of different system can be realized in a limited way only.

2.4   Need for a New Methodology and Requirements

After having carefully searched for selected approaches within the literature, it was found that numerous different methods for evaluating the flexibility of production systems exists. A comparison of the considered approaches, methods and models verified that none of these procedures met the requirements of the demands in practice (see Fig. 2.8).

Figure 2.8 clearly shows the lack of a procedure that enables a sufficient evaluation of Volume-, Mix-, and Expansion flexibility. Although some of these methods evaluate at least two of the required types of flexibility, they do show other deficits. Thus, the procedure represented by Schuh, Gulden, Wernhöner and Kamper enables

Degree of coverage: ◆ complete ◒ in parts ◇ not covered	Level				Dimension			Kind of flexibility			Intersectoral applicability	Independence of subjective evaluations	Comparability
	Factory	Segment	Line	Workplace	Costs	Time	Variety	Quantity	Product mix	Expansion			
PLANTCALC™	◆	◒	◒	◒	◆	◒	◆	◒	◒	◒	◆	◒	◆
ALI, SEIFODDINI	◒	◒	◆	◆	◇	◒	◆	◒	◆	◇	◇	◇	◒
HALLER	◆	◆	◆	◆	◆	◒	◆	◒	◒	◒	◒	◆	◇
WAHAB, WU, LEE	◇	◇	◇	◆	◆	◆	◆	◆	◆	◇	◆	◆	◆
SCHUH, GULDEN, WERNHÖNER, KAMPER	◒	◒	◆	◆	◆	◆	◆	◆	◆	◒	◆	◒	◆
DESYMA	◒	◒	◆	◆	◆	◒	◆	◒	◒	◆	◆	◒	◇
PELÁEZ-IBARRONDO, RUIZ-MERCADER	◇	◇	◆	◆	◆	◒	◆	◒	◒	◇	◒	◆	◇
FLEXIMAC	◆	◆	◆	◆	◇	◒	◆	◒	◒	◇	◆	◆	◒
PENALTY of CHANGE	◆	◆	◆	◆	◆	◒	◆	◒	◒	◆	◆	◒	◇
ZÄH, MÜLLER	◆	◒	◒	◒	◆	◆	◆	◆	◒	◒	◆	◆	◆
Demand of praxis	◆	◆	◆	◆	◆	◆	◆	◆	◆	◆	◆	◆	◆

Fig. 2.8 Qualitative classification of selected approaches for measuring flexibility in regard to praxis demands

evaluations of Mix- as well as Volume flexibility while also establishing indirect information on Expansion flexibility. However, they are not independent of subjective appraisal and the flexibility analyses are limited to the organizational levels workplace and line. The review of the consideration of different flexibility dimensions in the analyzed approaches emphasizes that all of them cover variety. All but two of the approaches at least partially consider all three dimensions. Furthermore, four procedures were identified which matched the multidimensional character of the flexibility at least partially, neglecting however, other aspects. The same applied for the applicability of some procedures in regard to organizational levels that had been defined in this book. Three of them corresponded completely to these requirements. Thus, the method "Penalty of Change" suited the application at a workplace- or even a factory level. On the other hand its downfall is that it disregards the time dimension and information about Quantity- or Product Mix flexibility, which cannot be concretely quantified. It is obvious that the evaluation ability of Volume-, Mix-, and Expansion flexibility seems to decrease, resulting in an increased level of consideration of all parameters in a production system.

Altogether one can say that different methods and procedures for measuring flexibility of production systems do exist (see Sect. 6.2). Many of them have a certain practical function and have been developed for specific purposes. That is why they are limited to isolated considerations. In contrast, procedures with a general structure, like the procedure of Zäh and Müller or PLANTCALC^{TM} enable more comprehensive considerations of flexibility. They are however not suited to detailed analyses at a line- or even workplace level. Furthermore, different procedures base themselves on a non-uniform terminology with regard to the interpretation of a particular kind of flexibility. Thus, Peláez-Ibarrondo and Ruiz-Mercader have a different understanding of Quantity- and Mix flexibility than Schuh, Gulden, Wernhöner and Kamper. While some of them consider these kinds of flexibility to be an auxiliary number for evaluating operational flexibility, the others consider it as a characteristic that has to be analyzed separately. In addition, different procedures can be observed whenever costs and time are being involved which differ in their regarded aspects and importance. The DESYMA-method for example, measures flexibility on the basis of application and efficiency, whereas Haller also involves qualitative elements.

Consequently, the usefulness of the application of all the procedures presented depends largely on the particular problem. Being more or less limited, they only partially fulfill the requests of the seeked evaluation method. A combination of different procedures that are considered to be adequate in order to design a comprehensive model does not seem to be useful, as each method provides different information and has been designed for different purposes and is based on different terminologies. In practice, this could result in non-transparent and even contradictory flexibility related results, which leads to the need for a new procedure design. Notwithstanding, particular approaches contain important aspects for evaluating flexibility of production systems, and should be a part of the evaluation method that has to be designed. Thus, Peláez-Ibarrondo and Ruiz-Mercader designate an interesting approach for evaluating Volume flexibility with the Break-even-Point. On the other side, Wahab, Wu and Lee as well as Schuh, Gulden, Wernhöner and Kamper also feature useful aspects regarding the Mix flexibility. Furthermore, these authors mention an interesting system concept for dynamic flexibility considerations.

2.5 Requirements for a New Evaluation Methodology

As a result of the identified flexibility challenges found in practice and according to analyses of the regarded domain, obvious weak points of existing evaluation methods were made apparent. That is why this chapter describes those demands that need to be satisfied in order to gain a goal orientated methodology for flexibility. Figure 2.9 summarizes the single demands in four groups that will be further explained in this chapter.

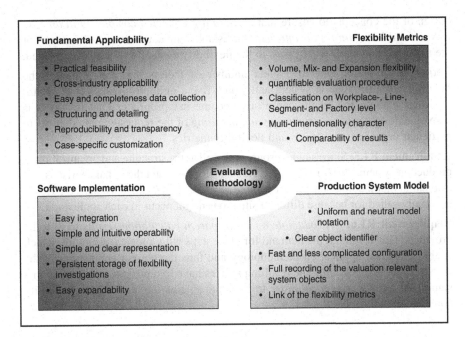

Fig. 2.9 Demands on the evaluation methodology

2.5.1 Fundamental Applicability

The evaluation methodology has to enable the production planner and manager to evaluate the flexibility of their production system in an easy way in order to draw meaningful conclusions on any adaptations and changes which might be necessary. This leads to the following requirements:

Requirement R1.1: A necessary criterion for the success of the application and the acceptance of a methodology is its *practical feasibility*, which is why its design needs to be orientated on practical flexibility challenges. Furthermore, flexibility has to be analyzed in a differentiated way on different organizational levels of production systems.

Requirement R1.2: In order to achieve a methodology that is propagated in practice and that enables advancements by means of analogies and comparisons of different production systems, a *cross-industry applicability* is of essential importance. Accordingly, flexibility evaluations should involve more than just isolated considerations; they should also be applicable universally and independently, without focusing on a certain scenario.

Requirement R1.3: In order to assure minimal effort in the execution of the evaluation methodology and in order to establish a great user acceptance, simple and comprehensive data collection is required. That is why care must be taken in the design of the concepts to ensure that all *data are collected completely in an easy way without losing any of the information*, Establishing data that are unnecessary or are not important at all with regard to the flexibility evaluations has to be avoided.

Requirement R1.4: Due to a great number of factors that can influence the flexibility of a production system and which risk a multiple overlap or disregard of different facts, the results can easily become inappropriate. That is why it is unavoidable to choose a *structuring and detailing* of the requested data that fits to particular organizational levels and flexibility metrics.

Requirement R1.5: Since single resources as well as the structure and organization of production systems differ in regard to their specification and their characteristics, the evaluation methodology has to be dyn*amic and case-specifically customized* . This is especially called for because different situations might occur in each application.

Requirement R1.6: *Reproducible and transparent calculation results* have to be provided. This results in the demand for clear steps with a logical structure, which enable a gradual application methodology and thus the correct application of the comprehensive procedure. The resulting indices have to represent the situation in a significant and clear way, in order to be applicable as the basis of the flexibility comparison and decision.

2.5.2   Flexibility Metrics

The development of flexibility metrics represents the main task concerning the design and realization of the evaluation methodology. Accordingly, this also ties up with the significant requirements, the fulfillment of which plays a major role in achieving the demanded purpose:

Requirement R2.1: In this book, decisive criteria for metrics are the evaluations of *Mix-, Quantity- and Expansion flexibility* as the preferred types of flexibility. This should provide information on how flexibly a production system can react to demand fluctuations with regards to quantity and product/variant mix and to what extent capacitive expansions are possible.

Requirement R2.2: In order to easily relate flexibility deficits to their responsible system objects, a focused analysis on the level of *workplace, line, segment* and *factory* is necessary.

Requirement R2.3: In order to represent the flexibility, a *quantifiable evaluation procedure* is required that, depending on the kind of flexibility and avoiding any subjective factors, calculates a specific index for each object that has to be evaluated. This guarantees a high degree of objectiveness which is the prerequisite for a fast evaluation of the actual flexibility potential that is unaltered when several data acquisitions occur. Furthermore, it helps to decide whether new flexibility potentials have to be created or if the existing flexibility is sufficient.

Requirement R2.4: In order to relate to the *multi-dimensional character* of flexibility, the applicability of the calculation of indices has to be assured in the context of a single-product as well as a multi-product production. Moreover, in addition to the variety, the corresponding time- and costs aspects also have to be adequately considered at the same time. In doing so, inference on an economical production must be possible in order to evaluate a company's economical balance between existing insecurities and its own flexibility (see Fig. 1.2., p. 6).

Requirement R2.5: The last metric-related request refers to a *comparability of the results*. On the one hand it includes the possibility to compare flexibility indices of different operation possibilities or between production systems with different dimensions in a branch. On the other, comparisons between production systems of different branches should be made possible, in order to conclude flexibility deficits of branch external systems on the basis of established flexibility parameters.

2.5.3 Production System Model

Requests on a production system model are derived from their underlying task which is the schematic demonstration of relations and dependencies of objectives that are relevant for the evaluation (workplace, line, segment and factory), as well as the linking of these with the flexibility metrics:

Requirement R3.1: In order to describe production systems which have to be analyzed in abstraction to the real world, a *uniform and neutral model notation* has to be chosen.

Requirement R3.2: In order to be able to represent all hierarchical and output related dependencies in a production system, a *full recording of the evaluation relevant system objects and their allocation relations* is requested. This should be carried out with an adequate degree of detail that conforms to the applicability of the flexibility metrics on different organizational levels of a production system.

Requirement R3.3: *Unique object identifiers* which assure the right connection between a represented symbol and the corresponding real system object are of extraordinary importance. Hereby inaccurate results due to incorrect or even contradictory data correlation are avoided.

Requirement R3.4: In order to guarantee a great user acceptance, a *fast and less complicated configuration* of the production system model is required, independent of any specific branch of industry. This demands a relatively high freedom with regard to the parametrization of different system objects.

Requirement R3.5: The last main criterion regarding a production system model results from the necessity to allocate distinct, corresponding flexibility indices to every single system object. For this purpose a *link of the flexibility metrics* to the production system is necessary, in order to provide different flexibility- and calculation information in relation to objects on different organizational levels of a production system.

2.5.4 Software Implementation

In the interest of having as little effort possible when using the evaluation methodology, its conversion to a software tool is imperative. This leads to the following requirements:

Requirement R4.1: The software has to be capable of an *easy integration* within the IT-infrastructure of the production. For this purpose, suitable interfaces for production related systems like Production Planning and Control (PPC) or Enterprise Resource Planning (ERP) systems and also for digital factory planning tools have to be provided, so that a great part of the necessary data can be accessed automatically.

Requirement R4.2: In order to provide a great acceptance by potential users in a company, the software tool should be *simple and intuitively operable,* without requiring comprehensive training courses in advance.

Requirement R4.3: There is the need for a *simple and clear representation* of evaluation objects, so that flexibility evaluations of a production system are easy to comprehend for the user. Thus, weak points have to be identified quickly and alternatives for their removal have to be represented.

Requirement R4.4: The *persistent storage of flexibility investigations* has to be guaranteed. It must be assured that the actual analysis status of a production system can be saved and re-accessed at any time. If needed, the comparison of different system alternatives and evaluations of different production systems models must be possible.

Requirement R4.5: An *easy expandability* of the software tool is a fundamental condition to keep to a minimum the effort for maintenance operations and subsequent expansions, which could result from the productive operation. For this purpose, a modular design of the software is recommended.

Chapter 3
Introduction of a New Evaluation Methodology

In the previous chapters dealt with the foundation of the Evaluation Methodology. Flexibility Indices in the context of this book are to be considered fundamentally as measured values, which use the quantifiable and reproducible measurement of relevant characteristics to define the flexibility of a production system. For the required calculation of these characteristics, special Flexibility Evaluation methods and metrics are necessary. The conceptual composition of these methods and their integration in the production system model are the subject of this chapter. Firstly, the fundamental approach to the planned calculation of the Volume-, Mix- and Expansion flexibilities should be defined and the parameters used to quantify them should be clarified. This results in the detailed definition of the calculation procedure for the measurement of the individual flexibilities. Due to their production-economic background, a special cost calculating reference frame will be defined, which will ensure the required quality of the calculated Flexibility Indices in that it guarantees a thorough cost record incorporating every level of observation. The presented approach to execution of a production system model illustrates how a data-specific connection can be made between a real world object and the Flexibility Evaluation method.

3.1 Fundamental Approach for the Evaluation of Volume-, Mix- and Expansion Flexibility

The types of flexibility defined in Sect. 2.2.3 lead to the conception of the following three proposals, which lay the foundation for the future implementation of the evaluation of the Volume-, Mix- and Expansion flexibilities in production systems. The primary aim is to record the quantifiable parameters of each evaluation method as well as to highlight which basic conditions the evaluation procedure needs to conform to.

S. Rogalski, *Flexibility Measurement in Production Systems*,
DOI 10.1007/978-3-642-18117-7_3, © Springer-Verlag Berlin Heidelberg 2011

3.1.1 Evaluation Proposal for Volume Flexibility

Drawing on the underlying understanding of Volume flexibility, a definition can be formed by means of a short term and at the same time economical adaptation of capacity, without changing the number of components or the structure of the production system in question (see Sect. 2.2.3.1). For this reason, the evaluation of this type of flexibility should be done by considering the profitability limits of the output achieved by the system within a specified time period (e.g., week or month). Two parameters are suitable in the quantifying of these limits of a production system:

- *Break-even-Point*, is the production quantity at which the revenues cover the total costs (variable and fixed costs), such that the resulting profit is zero.
- *Maximum Capacity*, indicates the biggest possible, while still profitable, output that a system is able to reach. It is determined by the technical performance of the production facility and also organizational measures like flexible man-hour models (e.g., overtime or shift work).

The Break-even-Point and Maximum Capacity cover an area in which the production output may vary, but which is still economical (see Fig. 3.1). This area, henceforth called the *Flexibility Range,* represents the Volume flexibility. This is expressed as a relative figure rather than an absolute one, so as not to distort the comparison between production systems with different dimensions. This is illustrated in the following example:

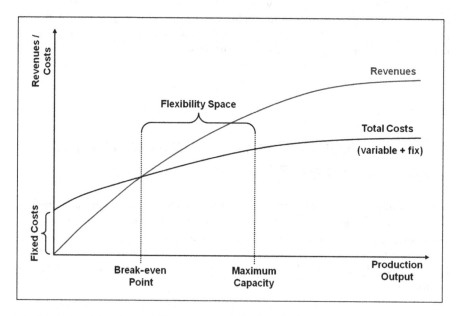

Fig. 3.1 Determining the flexibility range for a defined production programme

Example. Production system *A* exhibits an economic area (Flexibility Range) of 10,000–15,000 quantity units (QU), while Production system *B* has one of 1,000–5,000 QU. The Flexibility Range of *A* is therefore 1,000 QU bigger than that of *B*. Despite this fact, system *B* can be viewed as being more flexible since it allows a relative unit variation of 80%, whereas system *A* only allows 50%.

As a result of this method of considering the Volume flexibility, the following statement is valid:

The bigger the relative deviation between the Break-even-Point and the Maximum Capacity of a production system, the higher the Volume flexibility.

3.1.2 Approach to Evaluation of the Mix Flexibility

Mix flexibility, according to the previously introduced definition (Sect. 2.2.3.2), exhibits the freedom with the configuration of the production programme to be able to dispense of some products or to switch them with others, without influencing the system's optimal production profit. In order to make such a characteristic of a production system quantifiable, it is recommended to consider the difference in profit between the most lucrative product/variant configuration and those configurations which differ from it. The parameters which are necessary for this are:

- *System-optimal production profit*, is the biggest possible profit that a system can achieve with an optimal configuration of its production programme and based on a prescribed time line. This takes into consideration the type as well as the volume of the products to be manufactured.
- *Product-restricted optimum profit*, this indicates the maximum attainable profit of a system with the restriction of excluding one of the products which is usually manufactured onsite.

System-optimal production profit allows for the calculation (for each of the system's products) of the maximum profit which a system can still achieve if, under otherwise identical base conditions, one of the products of that system remains excluded from the production programme. The system-based Mix flexibility, expressed as a ratio for ease of comparison between differing production systems, is determined by using the so-called *Average production profit deviation*. This lends the evaluation a value for the extent of the affect that the non-production of individual goods has on the System-optimal production profit. For this it is advisable to calculate the Root Mean Square of the System-optimal production profit. This is a stricter assessment of "out-lying" values as opposed to a pure Arithmetic Mean method. Compared to the Standard Deviation, this method has the advantage of not being composed of the Arithmetic Means of the individual deviations. This leads to a more comprehensive detection of any threats to the success of the production and guarantees a realistic observation of the actual system behaviour when faced with changes in the production mix. Figure 3.2 illustrates the concept of the production profit deviation for determining the Mix flexibility.

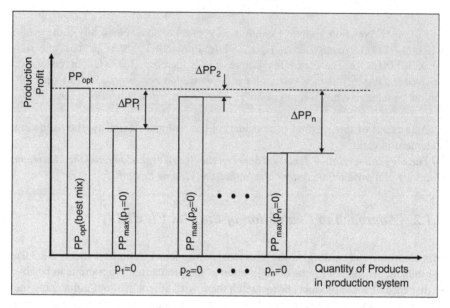

Fig. 3.2 Concept of the production profit deviation for determining the mix flexibility

As a result of this concept, a production system is *completely product-mix flexible* if the profit always remains constant independent of the selection of products to be manufactured. Based on this approach, the following statement for Mix flexibility is valid:

The lower the average production profit deviation of a system, the higher its Mix flexibility. The Mix flexibility is therefore optimal when all of the product-specific profit deviations calculated for the system are zero.

Since this type of approach is based solely on the evaluation of multiple-product manufacturing, it should not be utilised in the evaluation of single-product manufacturing. A system which is configured for the production of only one product/ variant is therefore defined as inflexible in terms of the product mix.

3.1.3 Evaluation Approach to Expansion Flexibility

Expansion flexibility describes the ability of a production system to sustainable increase its capacity through the changing of its structure and/or number of parts. A deciding criterion for this is the required investment for any additional expansion arrangements, which is why it becomes a pure cost consideration. This does however have the result that comparisons between evaluations of production systems of different size and capacity could be distorted. For this reason the definition of Expansion flexibility (as introduced in Sect. 2.2.3.3) uses the *economic expenditure* formulation. This should expresses the system specific Cost-Benefit-Relationship with which the expansion is possible. The difficulty lies in quantifying this relationship and with it the economic expenditure. Additionally, most of the various existing expansion alternatives exhibit differing time and cost expenditures and furthermore, lead to different capacity escalations. In order to still be able to quantify the economic expenditure, it is advised to use the following parameters:

- *Target capacity*, represents the benefit of the expansion and is to be seen as a fixed, user-dependant limit which indicates the volume-based expansion based on the existing maximum capacity.
- *Alternative-specific Break-even-point*, expresses, as a special characteristic of a potential expansions alternative, the new production volume of a system such that the resulting profit is able to cover the costs of the expansion.

Setting a prescribed Target capacity (e.g., +30%), has the advantage of eliminating any expansion alternatives which do not meet the new capacity requirements. All other alternatives which meet or exceed the Target capacity requirements are quantified based on a common evaluation principle. For this purpose, it is necessary to revert to the volume-based Flexibility range described in Chap 3.1.1. A fixed value is hereby presented by. In this way, the predefined Target capacity presents a fixed value which indicates whether any of the potential expansion alternatives overshoot the capacity limits of the system. This method avoids the development of unnecessary Flexibility alternatives, since the consequence is that the so-called *expansion-based Flexibility range*[1] is dependent on the Break-even-Point of the individual expansion alternatives (see Fig. 3.3). The closer an alternative comes to reaching this point, the larger its expansion-based Flexibility range and therefore the bigger its influence on the Cost-Benefit ratio of the system. In order to finally determine the economic expenditure for the production system, the expansion-based Flexibility range of the alternative with the best Cost-Benefit ratio (in relation to the *original Flexibility range*[2] of the system) is applied. The size of the resulting deviation between the two Flexibility ranges can then give information about the Expansion flexibility.

Based on this evaluation method, the following statement holds:

The Expansion flexibility of a production system is dependent on the expansion alternative with the highest Cost-Benefit ratio. The bigger the relative deviation between its Break-even-Point and its Target capacity, the more Expansion flexible the system.

Regarding the size change of the expansion-based Flexibility range, the following three cases need to be distinguished:

1. *The expansion-based Flexibility range is smaller than the original Flexibility range*
   The implementation of the expansion leads to a deterioration of Volume flexibility. A production system for which this is applicable should therefore be classified as "not completely Expansions flexible".

---

[1]The expansion related Flexibility range is determined by the break-even point of an expansion alternative and the specified target capacity. The maximum capacity, which in some cases may exceed the target capacity, is not considered.

[2]The original Flexibility range of a system is the area, generated within the limits of the Volume flexibility, between the Break-even point and the Maximum capacity (see Figure 3.1, p. 48). The current system status is considered without any expansion.

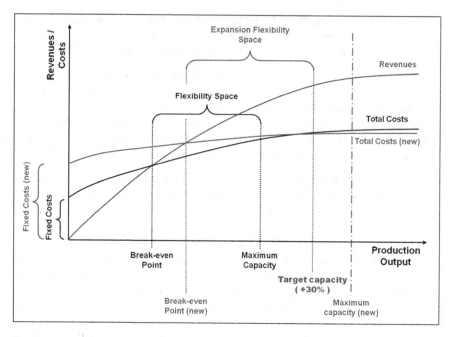

Fig. 3.3 Determining the cost-benefit ratio of an expansion alternative

2. *The expansion-based Flexibility range has the same size or is bigger than the original Flexibility range*
   The Volume flexibility remains unchanged or is even improved as a consequence of the expansion. For this reason, such a production system is considered to be "completely Expansion flexible".

3. *The Break-even-Point of the expansion-based Flexibility range is smaller than the Break-even-Point of the original Flexibility range*
   This implies that on the one hand, a "completely Expansion flexible" system as in case 2 is to be considered since its expansion-based Flexibility range is bigger than the original Flexibility range. However, the decreased Break-even-Point indicates that there is clearly optimization potential for the production system.

3.1.4 Comparative Consideration and General Conditions of the Evaluation

The conceptual approach to the calculation of those Flexibility types which are deemed relevant has been introduced. At this stage their significant features, taking into account the Flexibility dimensions and their calculation prerequisites, should be summed up.

The calculation of the Flexibility range is of central importance in the evaluation of **Volume flexibility**, since it gives information on the adaptation complexity regarding the handling of different production quantities. It thereby represents the variety dimension. In contrast, the cost dimension is considered in both limits of the Flexibility range. Their common characteristic is that the costs cannot exceed the revenues in this area and economical production can be assumed. The Break-even-Point indicates the left hand limit and is highly dependent on the attainable production-based turn over,[3] in which aspects of time also play a role. An example of this would be the Staff costs which are tied to the production times. On the other hand, the right hand limit of the Flexibility range is defined by the Maximum capacity, which is determined, above all, by the respective bottlenecks of the production process. It corresponds with the organizational operations options, like e.g.,) overtime, shift work or staff displacement and also the technical capacity of the production facility. In this relation the time dimension is especially relevant, since the production capacities are time dependent e.g.,) processing times of the product and operational hours of the facility. In the case of multi-product manufacturing, another factor can further influence the size of the Flexibility range, namely the composition of the product mix. As a result a change in the product mix particularly affects the Break-even-Point and the maximum capacity if – with a large, heterogeneous product allocation – there is a large deviation between the significant production parameters (e.g.,) set-up times, processing times) and the profit margin of the product. Therefore appropriate knowledge on the *composition of the product mix* and *the product-based sales price* is a necessary requirement in the evaluation of Volume flexibility.

Similarly, the concept of evaluation of the ***Mix flexibility*** incorporates all three flexibility dimensions. The variety dimension is hereby reflected in the average deviation of the production profit. This value shows the extent of the possible configurations for the composition of the product mix which results from the manufacturing spectrum of a production system. Provided that this is based on economic considerations as opposed to attainable production profit, then the cost dimension becomes especially relevant and makes the estimation of the economic risk of product mix fluctuations possible. The relevant factors in the average profit are on the one hand the corresponding product-dependent manufacturing costs which consist of the variable and fixed costs, and on the other hand the individual revenues of the products. The latter implies a *known sale price*. As in the determination of the Volume flexibility, temporal aspects are also closely linked to the cost dimension. The more resource time (e.g., staff and machine time) accumulated for the manufacture of a product, the higher the production costs. Therefore the third flexibility dimension, time, is also considered. The conceptual approach to the calculation of those Flexibility types which are deemed relevant has been introduced. At this stage their significant features, taking into account the Flexibility dimensions and their calculation prerequisites, should be summed up.

---

[3]The product based profit consists of the sales price of each product and the corresponding total costs (variable and fixed costs).

Borrowing from the calculation concept for the Volume flexibility, the evaluation of the **Expansion flexibility** will also take all three dimensions into account. The variety appears here in the expansion-based flexibility range, which explores the economic expenditure for the structural and component alignment of a production system which is needed to reach an ever-increasing capacity. For this purpose the Cost- Benefit ratio with reference to the cost dimension is defined for each allowable expansion alternative. This approach is similar to that of the Volume flexibility, with the so-called *Target capacity* replacing the Maximum capacity. This capacity represents a user-defined production requirement and is inevitably limited by the bottlenecks in the production process. In the case of multi-product production, this approach calls for a prescribed product mix ratio in order to be able to evaluate potential expansion alternatives. Furthermore, it is necessary that the sales prices of the products which make up the manufacturing spectrum are known. These prices in conjunction with the total costs (fixed and variable) determine the left hand limit of the Flexibility range (Break-even-Point). The inclusion of the time dimension within this calculation concept is carried out in different ways. On the one hand, it is analogous to the Quantity and Mix flexibility and corresponds with the cost dimension in that the necessary resource times are expressed as time. Examples of this would be the accrued working hours during the implementation of an expansion alternative, or the resultant change in manufacturing time. On the other hand, the time should also be interpreted as a definable frame for the execution of the expansion measures. As opposed to the Quantity and Mix flexibility, the underlying consideration here is a long term one.

Table 3.1 summarises once again, the relevant aspects of the three calculation concepts as presented here.

Table 3.1 Summarised overview of the important attributes of the calculation concepts

	Volume flexibility	Mix flexibility	Expansion flexibility
Describing (qualitative)	Ability of short term economical adaptation of capacity in case of fluctuations in demand	Economic risk of product mix fluctuations for the production profit	Ability to economical and sustainable increase of production capacity
Parameters (quantifiable)	– Break-even-point	– System-optimal production profit	– Alternative-specific break-even-point
	– Economic maximum capacity	– Product-restricted optimum profit	– Goal capacity
Dimensions	Costs, time, variety	Costs, time, variety	Costs, time, variety
Timeframe	Short-time	Short-time	Long-time
Requirements of calculation	– Known sales prices – Known product mix	– Known sales prices	– Known sales prices – Known product mix – Demand/goal of expansion

3.2   Calculation Procedure for the Measurement of Flexibility

Based on the same basic approach as previously described to assess the Volume-, Mix- and Expansion flexibility of production systems, the concept of the flexibility measurement draws upon quantifiable parameters. These are used to calculate specific key figures which are dependent on the Flexibility type. These parameters however, are often tied to time-varying as well as cost- and product related restrictions. They may therefore exhibit different fields of application due to the various possible manufacturing methods of one or more products, e.g.,) due to a number of existing identical or similar resources and production flows. This poses the risk of non-uniform or even conflicting Flexibility evaluations. It is therefore necessary to determine the respective parameters based on their limiting values (maxima or minima) which in turn can be traced back to a valid production schedule. This complies with the applicable restrictions, and so it is considered optimal. Such a production plan, often called a production program, describes the system resources in view of their nature and the extent of their application over a specific time period, which results in a defined number of products to be manufactured.

A fundamental commonality in the calculation of the various flexibilities within the evaluation methodology is therefore the determination of conforming, optimal production programmes, used to calculate the economic thresholds of the production. This results in optimisation problems being viewed in principle as linear. Their solution is usually achieved by using the simplex algorithm (see Sect. 6.1.2), regarded in practice as the most significant tool in solving linear optimization problems. Before this may be used, the optimization problem in question needs to be suitably formulated. This requires a mathematical model of the calculation parameters, result variables and mathematically logical relationships (auxiliary- as well as boundary conditions).

Before delving further into the calculations of the various flexibilities, a special basic algorithm should be defined to describe the standard, simplex-conforming optimisation problem, on which the evaluation methods of Volume-, Mix- and Expansion flexibility are based. The terms *Flexibility evaluation method* or *Flexibility metric* refer in particular to the summation of all the necessary calculations needed to determine a flexibility-type related index.

For better understanding of the calculation procedure described below within such an evaluation technique, the appropriate examples (that relate to the example presented in Sect. 6.4 of production system) are reverted to.

3.2.1   Base Algorithm

The Basic Algorithm itself consists of five consecutive calculative steps, which are formally described below and also illustrated by relevant, easily comprehensible examples.

3.2.1.1 Defining the Calculation Parameters (Step 1)

The existential condition for the utilization of the Basic Algorithm is the existence of the necessary calculation parameters, which reflect the flexibility-efficient characteristics of the production system. These parameters are to be understood as variable elements, which are equally valid for each production system, but which also assume different values due to the different characteristic values from system to system. The assignment of these values is in accordance with prescribed criteria and is usually possible through the automated or manual recourse to production related systems, such as Data Collection Systems (DCS) for the Production Planning and Control (PPC) or Enterprise Resource Planning (ERP).

In the table below are all the parameters necessary for the calculation of flexibility as well as their criteria for the value assignment. The three designated tables differ in terms of their non-cost-related parameters (Table 3.2), cost-related parameters (Table 3.3) and user-dependent parameters (Table 3.4).

Example. This example considers the two workplaces *WP1.1.1* und *WP 1.1.2* from the production system presented in the example in Sect. 6.4. They will receive the names WP_1 and WP_2.

According to the given parameters of the production system, three different types of products are produced at the workplaces: an intermediate product MI_1, a saleable intermediate product MI_2 and a finished product MI_3. WP_1 can produce both intermediate products MI_1 and MI_2, while WP_2 allows the production of MI_2 and the final product MI_3. This leads to the following workplace-product combinations:

$$R = \{(MI_1, WP_1), (MI_2, WP_1), (MI_2, WP_2), (MI_3, WP_2)\}$$

For the production of MI_3 however, two quantity units of MI_2 and one quantity unit of MI_1 are needed. It is important to note that the production of each MI_2 additionally requires one unit of MI_1. This results in the matrix of elements Table 3.5:

The sales price for the saleable intermediate product MI_2 amounts to 4 MU/QU and for the final product MI_3 to 11 MU/QU. In contrast, there is no market for the sale of intermediate product MI_1 the sales are thus assumed to be 0 MU/QU. The following revenue function for the example can be derived as:

$$s(x) = (0, 4, 11)^T$$

Furthermore, the parameters in Table 3.6 are given for the example of production system (see Sect. 6.4):

The sales ratio given for the production system ratio shows that twice as many MI_3 as MI_2 are to be produced, whereas MI_1 cannot be sold at all. It is therefore assumed to be 0 within this ratio, which results in Product mix vector:

$$v = (0 \times MI_1, \ 1 \times MI_2, \ 2 \times MI_3)^T = (0, 1, 2)^T$$

Table 3.2 Non-cost-related parameters

Parameter	Unit	Description/criteria for value assignment
FP	–	Quantity of finished products, where a final product is a manufactured item which has an existing market for its sale. It may also be part of a product which lies downstream in the production process, making it a so-called saleable intermediate-product
IP	–	Amount of intermediate products, where an intermediate product is a product manufactured by the production system itself, which is not saleable and is a component of a product downstream in the production process
M	–	Quantity of manufactured item with $M = \{MI_1, MI_2, MI_3, \ldots\}$, comprised of the quantity of finished products FP and the quantity of intermediate products IP
W	–	Quantity of workplaces $W = \{WP_1, WP_2, WP_3, \ldots\}$, used to manufacture the product/item
R	–	Ratio $R = \{r_1, r_2, r_3, \ldots\} \subset M \times W$ indicating whether a given product-workplace combination $r_i = (MI, WP)$ is valid; it can be a manufactured item MI produced at a workplace WP
WHM	–	Working-hour model applicable to the production system
$t_{\max}$	s	Maximum operating time available to the WHM under consideration
$t_{PT}(r_i)$	s	Process time for a product-workplace combination r_i, which indicates how much time within the chosen working-hour model WHM is needed to produce one manufactured item MI at workplace AP
$t_{aIT}(WP)$	s	Additional idle time (see Sect. 6.3.2), indicating the time period within the chosen working-hour model WHM in which workplace WP cannot produce due to disruptions
$a(r_i)$	%	Scrap rate for a product-workplace combination (MI, WP), within the considered working-hour model WHM
$s(MI)$	QU/MU	Sales price/proceeds for one manufactured item MI

s seconds, MU monetary unit, QU quantity unit, % percent

Table 3.3 Cost-related parameters

Parameter	Unit	Description/criteria for value assignment
$C_{\mathrm{var}}(r_i)$	QU/MU	Variable production costs of a product-workplace combination r_i for the chosen working-hour model WHM; they indicate how much it costs to produce one product unit MI at workplace WP (see Formula 3.38, p. 103)
C_{Fix}	MU	Fixed costs which are normalized to a period P and summed over all observation levels of a system which uses the working-hour model WHM (see Formula 3.39, p. 104)

MU monetary unit, QU quantity unit

Table 3.4 User-dependent parameters for calculation

Parameter	Unit	Description/criteria for value assignment
v	–	Vector for the identification of the desired product mix, is related to the desired volume ratio of the final product FP

Table 3.5 Element matrix from the example

	MI_1	MI_2	MI_3
MI_1	0	0	0
MI_2	1	0	0
MI_3	1	2	0

Table 3.6 Calculation parameters of the above-mentioned workplaces

System object	WP_1		WP_2	
E_{AP}	MI_1	MI_2	MI_2	MI_3
$C_{var}\ (MI, WP)$	0.50 MU	0.80 MU	1.00 MU	1.50 MU
a_{WP}	1%	2%	1%	5%
$t_{PT}\ (MI, WP)$	1 s	2 s	4 s	5 s
$t_{aIT,\ WP}$	4,500 s		3,000 s	
$C_{Fix}\ (M)$	250 MU		310 MU	
$t_{max}\ (WP)$	144,000 s			

3.2.1.2　Formulation of the Objective Function (Step 2)

By defining the necessary calculation parameters, the resulting variables are appointed in the form of a vector $x \in \mathbf{R}^{|R|+1}$. This describes a production plan, which indicates the production quantity $x(MI, WP)$ for each product $MI \in M$ and for each work plan $WP \in W$. As a last component it also contains an additional auxiliary variable, whose importance will be discussed in a subsequent chapter (see Formula 3.9, p. 63). In order to be able to calculate different resulting variables of the vector, an objective function is needed with which particular optimization problems regarding the determination of an optimal production plan can be solved. It can be represented by a vector $x \in \mathbf{R}^{|R|+1}$, where R declares the quantity of all possible product-workplace-combinations r. The target value of this function c, for a calculated production plan x, forms the scalar product of c and x according to Formula 3.1.

Formula 3.1 General formulation of the objective function

$$c(x) = c^T \cdot x = c_1 \cdot x_1 + c_2 \cdot x_2 + \cdots + c_{|R|} \cdot x_{|R|}$$

Every vector component c_i represents the target value that stands for the manufacture of a product MI at workplace WP. This objective function may vary depending on the procedure for calculating the Volume-, Mix- or Expansion flexibility (of the particular flexibility calculation method), which is caused by special criteria that determine the particular flexibility.

If the optimization's aim was to calculate a production plan that would maximize the total profit of the production system, the objective function from Formula 3.1

would have to be paraphrased, so that it names the amount of coverage for each vector component c_i, which is achieved by the production $r_i = (WP, MI)$ of a product MI at workplace WP. For that purpose, as demonstrated with Formula 3.2, the product sales price s_{MI} has to be multiplied with the rate of the effective production $(1 - a_{WP})$, which results in an average revenue. From this, component opportunity costs $C_{Opp}(MI)$ have to be subtracted, which result from the sum of sales of all products MI_j which are part of the product MI_i. On the other side, variable costs $C_{var}(MI, WP)$ of the corresponding product-workplace-combination r have to be calculated.

Formula 3.2 Calculation of the end value of a product related amount of coverage for a product-workplace-combination

$$c_i = (1 - a_{WP}) \cdot s_{MI} - C_{Opp}(MI) - C_{var}(r_j)$$

whereas:

$$r_i = (WP, MI);$$

$$C_{Opp}(MI) = \sum_{j=1}^{|P|} c_{i,j} \cdot e_j$$

According to Formula 3.2, the scalar product of the objective function vector c and a given production plan-vector x results in the entire coverage contribution of a wanted production plan. However, in order to gain the overall profit, fixed costs for the considered period P have to be subtracted from the entire coverage contribution, as demonstrated in Formula 3.3.

Formula 3.3 Objective function for optimizing the total gain

$$c(x) = c^T \cdot x - C_{Fix}$$

Example. The aim of the calculation example of the basic algorithm is to define a production plan that maximizes the achievable profit, under the inclusion of the given product mix, for the two workplace WP_1 and WP_2.

First of all, the production related amount of coverage per unit of quantity has to be calculated for each product-workplace-combination r, according to Formula 3.2.

For the product-workplace-combination (MI_3, WP_2) the following values can be calculated:

- Average sales price: $s'(MI_3, WP_2) = (1 - 5\%) \times 11MU = 10.45MU$
- Component opportunity costs: $C_{Opp}(MI_3) = 1 \times s(MI_1) + 2 \times s(MI_2) = 1 \times 0 + 2MU \times 4MU = 8MU$
- Variable costs: $C_{var}(EZG, AP) = 1.5\ MU$

Table 3.7 Basis data for calculating the objective function of the example

System Object	WP_1			WP_2		
M_{AP}	MI_1	MI_2	MI_3	MI_1	MI_2	MI_3
$s(MI)$ in MU/QU	0.00	4.00	–	–	4.00	11.00
$s'(MI, WP)$ in MU/QU	0.00	3.92	–	–	3.96	10.45
$C_{Opp}(MI)$ in MU/QU	0.00	0.00	–	–	0.00	8.00
$C_{var}(MI, WP)$ in MU	0.50	0.80	–	–	1.00	1.50
$c_{MI,\,WP}$ in GE/ME	**−0.50**	**3.12**	–	–	**2.96**	**0.95**

Consequently, the resulting corresponding amount of coverage for the objective function component $c_{MI_3,\,WP_2}$ is:

$$10.45MU - 8MU - 1.5MU = 0.95MU$$

The amounts of coverage of the other product-workplace-combinations can be gathered from the last line of the Table 3.7.

With the calculation of particular coverage amounts, the objective function can be established as follows (according to Formula 3.3):

$$c(x) = c_{1,1} \cdot x_{1,1} + c_{2,1} \cdot x_{2,1} + c_{2,2} \times x_{2,2} + c_{3,2} \times x_{3,2} - C_{Fix}$$
$$= 0.50 \cdot x_{1,1} + 3.12 \cdot x_{2,1} + 2.96 \cdot x_{2,2} + 0.95 \cdot x_{3,2} - (250 + 310)$$
$$= -0.50 \cdot x_{1,1} + 3.12 \cdot x_{2,1} + 2.96 \cdot x_{2,2} + 0.95 \cdot x_{3,2} - 560$$

Indication:
Due to reasons of simplicity, the indexing of (yet unknown) quantity proportions per product-workplace-combination occurs in the form of $x_{MI,WP}$ whereas *EZG* is the number of the product and *AP* is the number of the producing workplace. The same applies for the indexing of the components of the objective function.

3.2.1.3   Formulation of the Constraints (Step 3)

In order to correctly define the area of validity of flexibility related optimization problems, there is the need for constraints that determine mathematic-logic relations. These might however, just like the objective function, vary in dependence on the calculation of Volume-, Mix- or Expansion flexibility. Within the basis algorithm, it can be differentiated between two main groups of constraints that are also valid for the flexibility evaluation method. On the one hand, they concern the *time conditions* and on the other the *ratio conditions*.

Time Conditions

The first main group of constraints for flexibility related optimization problems are time conditions. They are based on the fact, that the production of a quantity unit of the product MI at workplace WP needs are certain process time $t_{PT}(MI, WP)$ (see Sect. 6.3.2). Furthermore, the maximal operation time for each workplace $t_{max}(WP)$ per period P is limited by the particular working time model WHM and the additional work plan specific holding time[4] $t_{aIT}(WP)$. This results in irregular conditions for each workplace within a production system, which are summarized in a matrix inequation according to Formula 3.4.

Formula 3.4 Time condition of the basis algorithm

$$\tilde{T}x \leq T_{max}$$

$\tilde{T}$ describes the matrix of parameter $|W| \times (|R| + 1)$. Each value $T_{i,j}$ of matrix $\tilde{T}$ stands for that process time, that workplace WP_i needs to produce product $r_{i,j}$. In the case that $r_{i,j}$ does not relate to workplace WP_i, the corresponding matrix shows the value 0 as the process time. In contrast, T_{max} represents the vector that indicates the maximal operation time $t_{max}(WP)$ less the occurring additional holding times $t_{aIT}(WP)$ for each workplace WP. Consequently, for calculating the real available operation time for a workplace, Formula 3.5 applies.

Formula 3.5 Calculation of the real available operation time of a workplace

$$T_{max} = [t_{max}(WP) - t_{aIT}(WP)]_{WP \in W}$$

According to inequation $\tilde{T}x \leq T_{max}$ the sum of the process times for all manufactured products at each workplace must not exceed the maximal available operation time, minus the workplace specific additional holding times.

Example. Time related inequation conditions shall be set up for the two workplaces WP_1 and WP_2, that have to be considered in an isolated way.

This demands, on the one hand, the calculation of their actual available operation time and on the other, the calculation of production dependent process times for each workplace, which can be determined on the basis of the parameters shown in Table 3.5 (p. 58) as follows:

- Real available operation time for WP_1: $T_{max,1} = 144{,}000s - 4{,}500s = 139{,}500s$
- Real available operation time for WP_2: $T_{max,2} = 144{,}000s - 3{,}000s = 141{,}000s$

---

[4]Provided that whenever workplaces are bonded (see Chap. 2.1.3.2), additional holding times occur, not all workplaces are being affected, only that one, that causes additional holding time. If the bonding-/transport system was responsible for additional holding time, they will be referred back to this workplace, that is positioned before the system. If several workplaces are concerned, an equal apportionment of the durability of the additional holding time on these workplaces occurs.

- Valid operation time for WP_1: $\tilde{T}_1 = 1s \cdot x_{1,1} + 2s \cdot x_{2,1}$
- Valid operation time for WP_2: $\tilde{T}_2 = 4s \cdot x_{2,2} + 5s \cdot x_{3,2}$

With the help of the resulting operation- and process times, the two time related inequation conditions can be formulated in the following way:

- $1s \cdot x_{1,1} + 2s \cdot x_{2,1} \leq 139,500s$
- $4s \cdot x_{2,2} + 2s \cdot x_{3,2} \leq 141,000s$

According to Formula 3.4, these inequation conditions have to be combined as a matrix inequation $\tilde{T}x \leq T_{max}$:

$$\begin{pmatrix} 1s & 2s & 0s & 0s & 0s \\ 0s & 0s & 4s & 5s & 0s \end{pmatrix} \leq \begin{pmatrix} 139,500 \\ 141,000 \end{pmatrix}$$

Indication:
    In its last column, matrix $\tilde{T}x$ (left part of the matrix inequation) does not relate to a possible product-workplace- combination r, but to the auxiliary variable that has been mentioned already (see Formula 3.9, p. 63).

Ratio Conditions

Ratio conditions, as the second basic group of constraints, result on the one hand from the given product mix for a flexibility calculations (see Table 3.1, p. 54); and on the other from component dependencies of single products. The latter can be described by a so-called component matrix $C \in \mathbb{R}^{|P| \times |P|}$. It shows, how many quantity units of a product are directly part of a quantity unit of another product (see Table 3.5, p. 58). Since no product can be a direct or indirect component of itself, the graph derived from C has to be cycle-free. The product mix itself that represents the quantity ratios of the final products with regard to their sale, is described by the vector $v \in \mathbb{R}^{|P|}$. For products for which no markets exist $MI_i \in IP$, $v_i = 0$ is valid. Due to this vector, production quantities of products which are destined to be sold and which are determined by the Simplex-Algorithm, have to be a multiple of v. This means, that the production quantity x'_i and x'_j of two sellable products $(MI_i, MI_j) \in M$ have to fulfill the condition $x'_i : x'_j = v_i : v_j$. However, since products manufactured by the same company can be part of other products according to the component matrix C and because scrap has to be taken into account, it is quite possible that the entire production quantity differs from this ratio.

In order to be able to express those ratio conditions, that differ from the product mix in a simple mathematical formula, a ratio matrix $V \in \mathbb{R}^{|P| \times (|R|+1)}$ that covers all product mix dependencies has to be defined. Furthermore, no quantity units that are unavailable due to scrap or further processing may be part of the ratio. On the one hand, this presumes the knowledge of each entire production quantity x_{MI_i} for each product MI_i and all workplaces WP. On the other, the total demand for each product $MI_i \in IP$ needed for the production of subsequent manufacture in the production process also has to be known. The basis algorithm calculates the production related entire production quantity x_{MI}, by (according to Formula 3.6) summing up the produced quantity $x(MI,WP)$ of the products MI, that are produced for each valid product-workplace-combination r at workplace WP, with the product-workplace-related scrap $a(MI, WP)$.

Formula 3.6 Calculation of the entire production quantity of a product

$$x_{MI} = \sum_{(MI, WP) \in R} (1 - a(MI, WP)) \cdot x(MI, WP)$$

The component matrix C is used for the calculation of the production related entire-intermediate-product quantity $x_{IP}(MI_i)$, i.e., the quantity of each product that is necessary for the production of other products from an entire production point of view. In doing so, the basis algorithm determines, with the help of data from $C_{i,j}$, the number of quantity parts that are necessary from MI_i to manufacture product MI_j. The given number has to be multiplied with the production quantity of product MI_j at each workplace WP, which is added up at all workplaces at the same time. Formula 3.7 describes this relation.

Formula 3.7 Calculation of the entire-intermediate-product-quantity of a production

$$x_{IP}(MI_i) = \sum_{(MI_j, WP) \in R} C_{i,j} \cdot x(MI_j, WP)$$

After having identified, with the help of the basis algorithm, the entire production quantity x_{MI_i} and the entire-intermediate-product-quantity $x_{IP}(MI_i)$ of a product, the quantity part y_i available for the sale of the corresponding product can finally be calculated. For this purpose, the entire-intermediate-production-quantity $x_{IP}(MI_i)$ has to be subtracted from the entire-production-quantity x_{MI_i} (see Formula 3.8).

Formula 3.8 Calculation of the product quantity that is available for the product mix

$$y_i = x_{MI_i} - x_{IP}(MI_i)$$

Since different quantity parts y_i have to maintain the ratio given in the product mix, the vector y is a multiple of production mix vector v. This ratio can be expressed with the auxiliary variable t, so that Formula 3.9 applies.

Formula 3.9 Characterization of quantity ratio with the help of the auxiliary variable "t"

$$y = t \cdot v$$

Therefore the following causal connection results for all products as represented in Formula 3.10:

Formula 3.10 Causal connection of quantity ratios

$$y_i = t \cdot v_i$$

$$\Leftrightarrow x_{MI_i} - x_{IP}(MI_i) = t \cdot v_i$$

$$\Leftrightarrow \left(\sum_{(MI_i, WP) \in R} a_{WP} \cdot x(MI_i, WP) \right) - \left(\sum_{(MI_i, WP) \in R} C_{i,j} \cdot x(MI_j, WP) \right) = t \cdot v_i$$

$$\Leftrightarrow \left(\sum_{(MI_i, WP) \in R} a_{WP} \cdot x(MI_i, WP) \right) - \left(\sum_{(MI_i, WP) \in R} C_{i,j} \cdot x(MI_j, WP) \right) - t \cdot v_i = 0$$

The left part of this equation can be considered as a scalar product between a vector $V_i \in \mathbb{R}^{|R|+1}$ and the vector of production plan x. This results in a ratio equation $V_i \cdot x = 0$ for each product EZG_i. With the basis algorithm these single ratio equations are summarized in a matrix equation according to Formula 3.11:

Formula 3.11 Ratio condition of the basis algorithm

$$V \cdot x = 0$$

With the help of this matrix equation an optimal production plan can be calculated. However, this plan does not have to totally utilize all production capacities, as the product mix also has to be regarded. Thus, the product that reaches its capacity border first is the limiting factor.

Example. In order to determine the ratio conditions derived from the product mix for workplace WP_1 and WP_2 of the example production system, considered in isolation, the relevant ratio matrix shall be set up.

According to Formula 3.8 the quantity of the products y_{MI_1}, y_{MI_2}, y_{MI_3} that are available for the product mix have to be identified. For the three products that were considered in the example, it consists of the following:

- Product MI_1 is produced at workplace WP_1 with a scrap of 1%. Thus, $0.99 \cdot x_{1,1}$ quantity units can be used for further production. Every produced MI_2 and every MI_3 use one piece of MI_1 each. That is why the corresponding number of $x_{2,1}$, $x_{2,2}$ and $x_{3,2}$ has to be subtracted from this. Consequently, $y_{MI_1} = 0.99 \cdot x_{1,1} - 1 \cdot (x_{2,1} + x_{2,2}) - 1 \cdot x_{3,2}$ quantity units remain for sale. As MI_1 is a non-sellable intermediate product and thus does not generate any sales, only that amount of quantity units shall be produced, that is necessary for the production of the other products. This results in the following equation:

$$0.99 \cdot x_{1,1} - 1 \cdot (x_{2,1} + x_{2,2}) - 1 \cdot x_{3,2} = 0$$

- Product MI_2 can be produced with the scape rates of 2% at workplace WP_1 and 1% at workplace WP_2. However, every produced MI_3 consumes 2 quantity units of EZG_2. That is why double the amount of the produced MI_3 has to be subtracted from the entire production quantity of MI_2. Consequently, the quantity of MI_2 remaining for sale is:

$$y_{MI_2} = 0.98 \cdot x_{2,1} + 0.99 \cdot x_{2,2} - 1 \cdot x_{3,2}$$

- As product MI_3 is not part of any other product, the amount of the quantity that is not sellable results directly from the scrap rate at the producing workplace WP_2. Therefore the following holds:

$$y_{MI_3} = 0.95 \cdot x_{3,2}$$

Based on the product mix of sale, that has to be followed in this calculation example, and which is described by vector $v = (0,1,2)^T$, the described ratio dependences can be summarized as follows:

$$0:1:2 = 0.99 \cdot x_{1,1} - 1 \cdot (x_{2,1} + x_{2,2}) - 1 \cdot x_{3,2} : 0.98 \cdot x_{2,1} + 0.99 \cdot x_{2,2} - 1 \cdot x_{3,2} : 0.95 \cdot x_{3,2}$$

By involving the auxiliary variable t, that has been introduced in Formula 3.9, this ratio coherence can be restated, in order to avoid complicated fractions. Thus, for a permitted production plan a value $t \in \mathbb{R}$ exists, for which it is imperative:

$$0 \cdot t = 0.99 \cdot x_{1,1} - 1 \cdot (x_{2,1} + x_{2,2}) - 1 \cdot x_{3,2}$$
$$1 \cdot t = 0.98 \cdot x_{2,1} + 0.99 \cdot x_{2,2} - 1 \cdot x_{3,2}$$
$$2 \cdot t = 0.95 \cdot x_{3,2}$$

By applying Formula 3.11, these coherences can be changed into the given matrix equation $V \cdot x = 0$, which happens as follows:

$$0 \cdot t = 0.99 \cdot x_{1,1} - 1 \cdot (x_{2,1} + x_{2,2}) - 1 \cdot x_{3,2}$$
$$1 \cdot t = 0.98 \cdot x_{2,1} + 0.99 \cdot x_{2,2} - 1 \cdot x_{3,2} \qquad \Leftrightarrow$$
$$2 \cdot t = 0.95 \cdot x_{3,2}$$

$$0.99 \cdot x_{1,1} - 1 \cdot (x_{2,1} + x_{2,2}) - 1 \cdot x_{3,2} - 0 \cdot t = 0$$
$$0.98 \cdot x_{2,1} + 0.99 \cdot x_{2,2} - 1 \cdot x_{3,2} - 1 \cdot t = 0$$
$$0.95 \cdot x_{3,2} - 2 \cdot t = 0$$

$$\Leftrightarrow \begin{pmatrix} 0.99 & -1 & -1 & -1 & 0 \\ 0 & 0.98 & 0.99 & -2 & -1 \\ 0 & 0 & 0 & 0.95 & -2 \end{pmatrix} \begin{pmatrix} x_{1,1} \\ x_{2,1} \\ x_{2,2} \\ x_{3,2} \\ t \end{pmatrix} = 0$$

$$\Leftrightarrow V \cdot x = 0 \text{ mit } V = \begin{pmatrix} 0.99 & -1 & -1 & -1 & 0 \\ 0 & 0.98 & 0.99 & -2 & -1 \\ 0 & 0 & 0 & 0.95 & -2 \end{pmatrix}$$

3.2.1.4 Formulation of the Linear Optimization Problem (Step 4)

According to the calculation steps carried out so far, a formulation of the optimization problem for the basis algorithm is possible in the following way:

A production plan $x \in \mathbb{R}^{|R|}$ is to be found, for which the objective function $c(x) = c^T x$ is maximised under the consideration of the following constraints:

- $\tilde{T}x \leq T_{max}$
- $V \cdot x = 0$
- $x \geq 0$ (nonnegative conditions)

As the Simplex-Standard-Solution procedure for linear optimization problems does not accept any linear equations but allows linear inequations as a secondary condition only; the linear equation conditions that have been formulated in the ratio matrix once again have to be restated. This is done by restating each ratio equation V_i of ratio matrix V in two inequations, each using the equivalence condition, because an equation $a = b$ is equivalent to $a \leq b \wedge (-a) \leq (-b)$. According to this Formula 3.12 applies for the ratio matrix.

Formula 3.12 Restating of the matrix equation to the matrix inequation

$$V \cdot x = 0$$
$$\Leftrightarrow V \cdot x \leq 0 \ \wedge \ V \cdot x \geq 0$$
$$\Leftrightarrow V \cdot x \leq 0 \ \wedge -V \cdot x \leq 0$$

Applying Formula 3.12 enables summarizing the constraints that apply for the basis algorithm, in a simplex adequate way, as demonstrated in Fig. 3.4.

Requirements	Find the production plan $x \in \mathbb{R}^{|R|}$, which leads to a maximum objective function!
Objective Function	$c(x) = c^T x$
Constraints	▪ $\tilde{T}x \leq T_{max}$ ▪ $V \cdot x \leq 0$ and $-V \cdot x \leq 0$ ▪ $x \geq 0$

Fig. 3.4 Simplex-adequate optimization problem for the basis algorithm

3.2.1.5 Solution of the Optimization Problem (Step 5)

Due to rephrasing the ratio conditions in more linear inequations, the mathematic model that applies for the basis algorithm, consisting of calculation parameters, resulting variables and constraints, now exists in a valid form that allows the application of the Simplex-Standard-Solution procedure, in order to establish an optimal production plan (see Sect. 6.1.2).

Example. On the basis of the calculated values for the two workplaces WP_1 and WP_2, which are being considered in isolation in the example production system, the optimization problem is formulated Table 3.8:

The application of the Simplex-Standard-Solution procedure in this optimization problem calculates the vector x as an optimal production plan for the two workplaces WP_1 and WP_2:

$$ x = \begin{pmatrix} x_{1,1} \\ x_{2,1} \\ x_{2,2} \\ x_{3,2} \\ t \end{pmatrix} = \begin{pmatrix} 67,297 \\ 36,103 \\ 11,587 \\ 18,930 \\ 8,992 \end{pmatrix} $$

This results in the following number of quantity units for workplace WP_1 that have to be produced:

- $MI_1 = 67,294\ \mathrm{QU}$
- $MI_2 = 36,103\ \mathrm{QU}$

Table 3.8 Simplex-adequate formulation of the optimization problem for the calculation example

Requirements	Find the production plan $x = (x_{1,1}, x_{2,1}, x_{2,2}, x_{3,2})^T$, which leads to a entire maximum profit for both workplaces WP_1 und WP_2!
Objective function	$c(x) = -0.50 \cdot x_{1,1} + 3.12 \cdot x_{2,1} + 2.96 \cdot x_{2,2} + 0.95 \cdot x_{3,2} - 560$
Constraints	$\blacksquare \begin{pmatrix} 1s & 2s & 0s & 0s & 0s \\ 0s & 0s & 4s & 5s & 0s \end{pmatrix} \leq \begin{pmatrix} 139,500 \\ 141,000 \end{pmatrix}$
	$\blacksquare \begin{pmatrix} 0.99 & -1 & -1 & -1 & 0 \\ 0 & 0.98 & 0.99 & -2 & -1 \\ 0 & 0 & 0 & 0.95 & -2 \end{pmatrix} \cdot \begin{pmatrix} x_{1,1} \\ x_{2,1} \\ x_{2,2} \\ x_{3,2} \\ t \end{pmatrix} \leq 0$
	$\blacksquare \begin{pmatrix} -0.99 & 1 & 1 & 1 & 0 \\ 0 & -0.98 & -0.99 & 2 & 1 \\ 0 & 0 & 0 & -0.95 & 2 \end{pmatrix} \cdot \begin{pmatrix} x_{1,1} \\ x_{2,1} \\ x_{2,2} \\ x_{3,2} \\ t \end{pmatrix} \leq 0$
	$\blacksquare\ x \geq 0$

In contrast, workplace WP_2 has to produce these quantity units:

- $MI_2 = 11,587\,\text{QU}$
- $MI_3 = 8,992\,\text{QU}$

Adhering to the given product mix results in an entire maximum profit for both workplaces (considered in isolation) of:

$$c(x) = 131,276\,\text{MU}$$

3.2.2 Flexibility Evaluation Method of Volume flexibility

According to the evaluation approach explained in Sect. 3.1.1, the maximum capacity, the break-even point and the resulting flexibility range have to be determined in order to quantify the Volume flexibility. The procedure for calculating these parameters is part of the following chapters. This needs to be provided for each system object (whole system or subsystems) of the organization hierarchy that has been defined in Sect. 2.1.3, from the factory via segment and line up to the point of the workplace.

3.2.2.1 Maximum Capacity

The maximum capacity characterizes the maximum attainable output of a system object within a valid working hour model, while considering a given product mix. It has to be determined for each evaluation relevant object by involving scrap rates as well as additional holding times and must not risk the efficiency of the total production. Accordingly, the task for the following calculations is to determine a production plan for each system object which, due to a given product mix, provides a preferably high output, so that the attainable profit of the whole system will not become negative. Here we must differentiate between a calculation procedure for the whole system and the corresponding subsystems within it. The reason for this is that the product mix required for the whole system, is not necessarily valid for the subsystems, as these do not unconditionally produce all of the required products on their own. As long as there are possibilities for compensations by other subsystems, discrepancies are acceptable.

Calculation of the Maximum Capacity for the Whole System

The calculation of the maximum capacity for the whole system is carried out as a linear optimization problem with the use of the basis algorithm that has been described before. However, this requires minor adaptations of the basis algorithm. Firstly it concerns the objective function c, which has to be converted to a constantly positive function, so that the product-workplace related output x_i that is represented by the production plan-vector x is maximised. That is why the specific objective values of c have to be used with +1 ($c_i = 1$), which is demonstrated in Formula 3.13.

Formula 3.13 Objective function for calculating the maximum capacity (constantly positive)

$$c(x) = x_1 + x_2 + \cdots + x_{|R|} \rightarrow \text{Max.}$$

Secondly, an additional secondary condition aside from the basis algorithm is needed, which guarantees that only those production programmes that do not result in a negative total profit are accepted as a solution, That is why the objective function for optimizing the total profit is used, as represented in Formula 3.3 (p. 59) and named $g(x)$ as a limiting profit function (see Formula 3.14) and which is applied within the constraints (see Fig. 4.4) is used.

Formula 3.14 Profit function as an additional secondary condition for the basis algorithm

$$g(x) = g^T \cdot x - C_{Fix}$$

By considering these two modifications of the basis algorithm the optimization problem for calculating the maximum capacity of the whole system can be demonstrated as shown in Fig. 3.5.

Example. When the calculation procedure is applied to the example production system that is represented in Sect. 6.4 as previously described, a period related maximum capacity $x_{MAX}(Factory)$ of 209,876 quantity units results for the whole system "factory". The connected production plan is presented in *line x* (*EZG, AP*) of the Table 3.9.

Requirements	Find the production plan $x \in \mathbb{R}^{|R|}$, which leads to a maximum objective function!
Objective Function	$c(x) = x_1 + x_2 + \cdots + x_{|R|}$
Constraints	• $g^T \cdot x - C_{Fix} \geq 0$ • $\tilde{T}x \leq T_{\max}$ • $V \cdot x \leq 0$ and $-V \cdot x \leq 0$ • $x \geq 0$

Fig. 3.5 Simplex-adequate optimization problem for calculating the maximum capacity for the whole system

Table 3.9 Production quantities for maximizing the capacity of the system object "factory" of the production system in the example

System object	WP1.1.1		WP1.1.2		WP1.0.1	WP2.1.1	WP2.1.2	WP2.1.3
M_{WP}	MI_1	MI_2	MI_2	MI_3	MI_4	MI_4	MI_5	MI_6
$x(MI, WP)$ in QU	41,851	29,680	0	11,752	10,326	70,000	34,875	11,392
$\sum x(WP)$ in QU	71,531		11,752		10,326	72,000	34,875	11,392
System object	$L1.1$				–	$L2.1$		
$\sum x(Linie)$ in QU	93,609				–	116,267		
System object	$S1$					$S2$		
$\sum x(Segment)$	93,609 QU					116,267 QU		
System object	Factory							
$x_{MAX}(Factory)$	**209,876 QU**							

Calculation of the Maximum Capacity for Subsystems

As mentioned before, it is not useful to adhere to the product mix of the whole system when considering the subsystems. That is why for subsystems specific objects or, if need be, a new product mix has to be defined. Starting with the production plan of the higher system, only the part relevant for the subsystem is considered. This contains the quantity ratios of those products, which the subsystem can also produce. Based on the results of the calculation in Table 3.9, the new product mix vector for segment $S1$ of the example production system in Sect. 6.4 can be consequently determined as explained in the following example:

Example. The super-ordinate system of $S1$ is the factory itself, whose product mix vector relates to $v = (0; 1; 2; 1.5; 0; 2)^T$. Calculating the maximum capacity of the system object "factory" leads to following production quantities for segment $S1$:

$$MI_1 = 41,851\,QU, MI_2 = 29,680\,QU, MI_3 = 11,752\,QU, MI_4 = 10,326\,QU$$

Production of the two products MI_5 and MI_6 in segment $S1$ is not possible, which is why the following number of items ratio applies for the segment:

$$MI_1 : MI_2 : MI_3 : MI_4 = 41,851 : 29,680 : 11,752 : 10,326$$

The new product mix vector that relates to segment $S1$ can be derived from this:

$$v = (3.56; 2.53; 1; 0.88)^T$$

On the one hand, the advantage of this procedure is the fact that every subsystem produces at least as much as is required by its super-ordinate system, in order to achieve its capacity limit, because the new product mix always allows that very product combination. On the other hand, additional available capacities can be used, if they exist, as the subsystem is no longer limited by other subsystems.

An example of this is a subsystem in the factory that is only able to deliver intermediate products to a certain extent. After having determined the new product mix for a subsystem, a specific linear optimization problem can be formulated for it, which is again based on the basis algorithm. However, two main aspects have to be considered:

- The objective function must correspond to the function used for calculating the maximum capacity for the whole system (see Formula 3.13, p. 69). However, it is only the involved product-workplace- related outputs x_i, which affect the system itself.
- Component dependencies can be disregarded, since the part of the intermediate products is already included in the new product mix. Accordingly, the new ratio conditions of the basis algorithm (see Sect. 3.2.1.3) have to be formulated without subtraction of the total intermediate product quantities $x_{IP}(MI)$. According to Formula 3.10 (p. 64) the following causal connection results:

Formula 3.15 Demonstration of the causal connection of ratio conditions for subsystems

$$V_i \cdot x_i = 0$$

$$\Leftrightarrow x_{MI_i} - t \cdot v_i = 0$$

$$\Leftrightarrow \left(\sum_{(MI_i, WP) \in R} a_{WP} \cdot x(MI_i, WP) \right) - t \cdot v_i = 0$$

In order to eliminate mix-ups with the original ratio conditions of the basis algorithms, the modified ratio equations $V_i \cdot x = 0$ have to be described as $\tilde{V}_i \cdot x = 0$. According to Formula 3.11 (p. 64) the matrix equation $\tilde{V} \cdot x = 0$ has to be transferred to the simplex- adequate notation as shown by Formula 3.16, with the help of transformation rules that have already been shown (see Formula 3.12, p. 66).

Formula 3.16 Ratio condition for calculating the maximum capacity of subsystems

$$\tilde{V} \cdot x \leq 0 \quad \text{and} - \tilde{V} \cdot x \leq 0$$

According to the adaptations of the basis algorithm, the optimization problem for calculating the maximum capacity for subsystems takes the form that is demonstrated in Fig. 3.6.

Example. Based on the explained calculation procedure for subsystems, the result for a maximum capacity $x_{MAX}(S_1)$ for segment S1 of the example production system (see Sect. 6.4) is 150,579 QU per period. Table 3.10 shows the corresponding production plan $x(MI, WP)$.

Requirements	Find the production plan $x \in \mathbb{R}^{|R|}$, which leads to a maximum objective function!
Objective Function	$c(x) = x_1 + x_2 + \cdots + x_{|R|}$
Constraints	■ $\tilde{T}x \leq T_{max}$ ■ $\tilde{V} \cdot x \leq 0$ and $-\tilde{V} \cdot x \leq 0$ ■ $x \geq 0$

Fig. 3.6 Simplex-adequate optimization problem for calculating the maximum capacity for subsystems

Table 3.10 Production quantities for maximization of the capacity of segment S1 in the example production system

System object	WP1.1.1		WP1.1.2		WP1.0.1
M_{WP}	MI_1	MI_2	MI_2	MI_3	MI_4
$x(MI, WP)$ in QU	67,374	36,062	11,600	18,919	16,624
$\sum x(WP)$ in QU	103,436		30,519		16,624
System object	L1.1				–
$\sum x(Linie)$ in QU	133,955				–
System object	S1				
$x_{MAX}(S_1)$ in QU	**150,579**				

3.2.2.2 Break-even Point

The aim of the calculation of the break-even point is to measure a minimum production quantity (break-even quantity) for a chosen system object within the valid working hour models. This quantity has to guarantee, under consideration of the given product mix and the causal coherences, that the sale returns correspond to the total costs of the production system. Thus, the achievable total profit has to be zero. Again, we differentiate between the calculation of the break-even point for subsystems and for the whole system.

Calculation of the Break-even Point for the Whole System

The calculation of the break-even amount for the whole system can be executed relatively easily, using the previously described basis algorithm. However, the objective function c that has to be maximized also has to be restated in terms of a constantly negative function, so that the production plan represented by vector x is kept to a minimum in regard to its product-work station-related outputs x_i. For this purpose, single objective values c_i receive the value -1 ($c_i = -1$), as shown with Formula 3.17.

Formula 3.17 Objective function for calculating the break-even point (constantly negative)

$$c(x) = -x_1 - x_2 - \cdots - x_{|R|} \rightarrow \text{Max.}$$

However, the application of this objective function in connection with the basis algorithm alone is not sufficient to determine the break-even point, because production plans are also valid, which can translate to economical disadvantages for the whole system. That is why another secondary condition has to be added, that allows only those production programmes that do not result in a negative total profit. For this purpose, the profit function $g(x) = g^T \cdot x - C_{Fix}$, that has also been used for calculating the maximum capacity for the whole system, is used (see Formula 3.14). Thus, the optimization problem for calculating the break-even point that has been demonstrated in Fig. 3.7 can be deduced for the whole system.

Example. The solution of the break-even optimization problem for the whole system factory that is mentioned in Sect. 6.4 is a quantity x_{BE} *(Factory)* of 152.278 quantity units per period that refers to the production plan in Table 3.11.

Requirements	Find the production plan $x \in \mathbb{R}^{|R|}$, which leads to a maximum objective function!
Objective Function	$c(x) = -x_1 - x_2 - \cdots - x_{|R|}$
Constraints	• $g^T \cdot x - C_{Fix} \geq 0$ • $\tilde{T}x \leq T_{max}$ • $V \cdot x \leq 0$ and $-V \cdot x \leq 0$ • $x \geq 0$

Fig. 3.7 Simplex-adequate optimization problem for calculating the break-even point for the whole system

Table 3.11 Production quantities that are involved in the calculation of the break-even point for the system object "factory" of the production system in the example

System object	WP1.1.1		WP1.1.2		WP1.0.1	WP2.1.1	WP2.1.2	WP2.1.3
M_{WP}	MI_1	MI_2	MI_2	MI_3	MI_4	MI_4	MI_5	MI_6
$x(MI, WP)$ in QU	30,335	21,513	0	8,518	0	58,377	25,278	8,257
$\sum x(WP)$ in QU	51,848		8,518		0	58,377	25,278	8,257
System object	L1.1					L2.1		
$\sum x(Linie)$ in QU	60,366					91,912		
System object	S1					S2		
$\sum x(Segment)$	60,366 QU					91,912 QU		
System object	Factory							
x_{BE}(Factory)	**152,278 QU**							

Calculation of Break-even Points for Subsystems

Determining break-even points for subsystems is more complicated than the determination for the whole system, as one cannot presume that all the products manufactured there are sellable. A subsystem could exclusively produce intermediate products for another subsystem, which cannot be sold. In such a case, according to the calculation of the break-even point for the whole system, the best solution would be not to produce intermediate products that are not sellable for the concerned subsystem. As the profit decreases with each of those intermediate products, the costs increase instead of the revenues. However, these considerations are not very helpful, as these intermediate products are necessary for other tasks of the production. Hence, another procedure is required in order to calculate Break-even quantities for subsystems. Two basic thoughts have to be included:

- If one subsystem manufactures one product in a more expensive way than another subsystem, it has negative consequences for its Break-even point. This is because, in accordance with the cost ratio, more products have to be fabricated than in the first subsystem.
- In order to determine such a cost ratio it is advisable to consider the average costs, because subsystems can fabricate the same products at different costs, as for example in a segment where different workplaces produce simultaneously.

According to these two thoughts, average costs have to be determined for all fabricated products in each subsystem. For this purpose the calculation of the maximum capacity for the whole system is required (see Fig. 3.5, p. 69), which allows us to revert to the representative quantity units that were used to calculate average costs. Each product-workplace-combination of a subsystem has a certain output for a certain product. According to Formula 3.18, this output has to be multiplied with the output related production costs, and then the outputs of all the product-workplace-combinations have to be added up. If a subsystem does not manufacture a product according to the production plan,[5] although it is theoretically possible, the variable costs of the relevant product-workplace-combination are used as the average costs. However, if there are several product-workplace-combinations, the arithmetic mean of the variable costs is used. The resulting subsystem related total costs of a product have to be divided by its total quantity afterwards.

---

[5]Here, it is referred to the production plan, which results from the calculation of the maximum capacity for the whole system.

Formula 3.18 Determining of product related average costs for a subsystem

$$K_{avg}(MI, S) = \frac{1}{n_{MI,S}} \cdot \sum_{WP \in S} n(MI, S) \cdot C_{var}(r_k),$$

whereas:

$C_{avg}(MI, S)$: Average costs of a product MI in subsystem S;
$n(MI, S)$: Produced number of a product MI in subsystem S;
$C_{var}(r_k)$: Variable production costs for a product-workplace-combination

Using all the calculated average costs, the subsystem related cost ratio can be calculated for each product on all system levels. For this purpose, different average costs of all subsystems of an organizational level have to be added up, and the sum has to be divided by each single subsystem specific value of the average costs (see Formula 3.19).

Formula 3.19 Calculation of product related cost ratio for subsystems

$$v_C(MI, S_i) = \frac{C_{avg}(MI, S_i)}{\sum\limits_{S \in L} C_{avg}(MI, S)},$$

whereas:

$C_{avg}(MI, S_i)$: Average costs of a product MI for subsystem S_i;
$v_C(MI, S_i)$: Cost ratio of a product MI for subsystem S_i;
L: Organizational level of the production system;
$\sum\limits_{S \in E} C_{avg}(MI, S)$: Sum of the average costs of all subsystem of organizational level L

After determining the different cost ratios for each subsystem, they must be incorporated in the optimization problem of the calculation of the Break-even point for the whole system (see Fig. 3.7, p. 73). Accordingly, this requires the extension of the ratio conditions which refer to the basis algorithm (see Formula 3.11, p. 64). Here, Formula 3.10 (p. 64) can be restated to give $x_{MI_i} = t \cdot v_i + x_{IP}(MI_i)$, which results in the total production quantity of a known product within the whole system. Correspondingly, the total production quantity of a certain product in a special subsystem can be determined as follows:

Formula 3.20 Calculating of the total production quantity of a product for a special subsystem

$$x_{MI,WP} = \sum_{(MI,WP) \in R, WP \in S} (1 - a(MI, WP)) \cdot x(MI, WP)$$

Based on Formula 3.20 and with the consideration of ratio $v_C(MI, S)$ that has been determined before, the following causal mathematical connection results:

Formula 3.21 Relationship that includes cost ratios in ratio conditions of the basis algorithm

$$x_{MI,WP} = v_C(MIG, S) \cdot x_{MI_i} = v_C(MI, S) \cdot (t \cdot v_i + x_{IP}(MI_i))$$

$$\Leftrightarrow \sum_{(MI, IP) \in R, WP \in S} (1 - a(MI, WP)) \cdot x(MI, WP)$$

$$= v_C(MI, S) \cdot \left(t \cdot v_i + \sum_{(MI_j, WP) \in R} C_{i,j} \cdot x(MI_j, WP) \right)$$

$$\Leftrightarrow \sum_{(MI, WP) \in R, WP \in S} (1 - a(MI, WP)) \cdot x(MI, WP)$$

$$- v_C(MI, S) \cdot \left(t \cdot v_i + \sum_{(MI_j, WP) \in R} C_{i,j} \cdot x(MI_j, WP) \right) = 0$$

The left part of Formula 3.21 can be written as the scalar product between a vector $V_i \in \mathbb{R}^{|R|+1}$ and the vector of production plan x. Hence, a ratio equation in the form $v \cdot x = 0$ follows for each subsystem and each product. Then, according to Formula 3.21, all resulting ratio equations can be summarized as a matrix equation:

Formula 3.22 Ratio condition for determining break-even points for subsystems

$$V_{BE} \cdot x = 0$$

The adapted ratio conditions then replace the existing ratio conditions of the optimization problem described.

Example. According to Table 3.9 (p. 70) a maximum capacity of 209,876 QU has been calculated for the production system in Sect. 6.4. This maximum capacity has the same value for each organizational level in the system. In order to obtain Break-even quantities of the two segments $S1$ and $S2$ from this, the average costs of all the manufactured products in each segment have to be calculated. The concrete values are represented in the following Table 3.12.

Here, product MI_4 is a particular case, because it can be produced in both segments. However, average costs in segment $S2$ are 0.70 MU/QU less. By applying Formula 3.18 (p. 75) and Formula 3.19 (p. 75) the following segment related production-cost-ratio can be calculated for MI_4:

Table 3.12 Product related average costs on factory level for the example production system

Product	MI_1	MI_2	MI_3	MI_4	MI_5	MI_6
$C_{avg}(MI, S_1)$ in MU/QU	0.50	0.80	1.50	2.00	0	0
$C_{avg}(MI, S_2)$ in MU/QU	0	0	0	1.30	0.90	1.70

Requirements	Find the production plan $x \in \mathbb{R}^{|R|}$, which leads to a maximum objective function!
Objective Function	$c(x) = -x_1 - x_2 - \cdots -x_{|R|}$
Constraints	<ul><li>$g^T \cdot x - C_{Fix} \geq 0$</li><li>$\tilde{T}x \leq T_{\max}$</li><li>$V_{BE} \cdot x \leq 0$ and $-V_{BE} \cdot x \leq 0$</li><li>$x \geq 0$</li></ul>

Fig. 3.8 Simplex-adequate optimization problem for calculating break-even points of subsystems

Table 3.13 Break-even quantities of subsystems for the production system in the example

System object	WP1.1.1	WP1.1.2	WP1.0.1	WP2.1.1	WP2.1.2	WP2.1.3
x_{BE} (WP) in QU	39,896	20,470	35,380	22,997	25,278	8,257
System object	L1.1		–	L2.1		
x_{BE} (Linie) in QU	60,366		–	56,532		
System object	S1			S2		
x_{BE} (Segment) in QU	95,746			56,532		

$$v_C(MI_4, S_1) : v_C(MI_4, S_2) = \frac{2}{3.3} : \frac{1.3}{3.3}$$

On the basis of this ratio, ratio conditions for the segment level have to be restated to give Formula 3.21 (p. 76), which will the result in the optimization problem of the break-even calculation for the subsystems (see Fig. 3.8, p. 77). Related to the break-even quantity of MI_4, the following segment dependent quantity division results:

- $(MI_4, S_1) = 35,380 \, \text{QU}$
- $(MI_4, S_2) = 22,997 \, \text{QU}$

This therefore leads to a total output for all manufactured products of 95,746 QU for Segment S1 and 56,532 QU for S2.

Table 3.13 shows the calculated break-even quantities x_{BE} for all subsystems of the system in Sect. 6.4 which were calculated using the above mentioned method.

3.2.2.3 Calculation of the Volume Flexibility

Previous calculations of the break-even quantity and maximal capacities of the whole system and its subsystems are the base for determining different Volume flexibilities. They allow for the calculation of the specific flexibility range for each system object. In order to do so, according to Formula 3.23, the smallest break-even quantity $x_{BE}(S)$ has to be subtracted from the greatest possible production quantity (maximum capacity) $x_{MAX}(S)$, from all working hours-models that apply for the object.

Formula 3.23 Calculation of the flexibility range for different system objects

$$x_{MAX}(S) - x_{BE}(S)$$

Hence, the flexibility range characterizes the dimension of the economical, object-dependent part of production, in the form of an absolute value. However, in order to be able to compare different system objects, this flexibility range has to be put in a relationship with the maximum capacity, as shown in Formula 3.24.

Formula 3.24 Calculation of the volume flexibility for different system objects

$$F_{Volume}(S) = \frac{x_{MAX}(S) - x_{BE}(S)}{x_{MAX}(S)} \cdot 100\%$$

Example. Applying Formula 3.24 to different system objects in the example production-system leads to the indices of Volume flexibility that are represented in Table 3.14.

Corresponding to the calculation of the Volume flexibility, and in accordance with the index for the system object "factory", the production system is able to compensate for approximately 27.5% fluctuation in demand (based on the capacity limit and for a given product mix $MI_2 : MI_3 : MI_4 : MI_6 = 1 : 2 : 1.5 : 2$), without endangering the economics and reliability of the production. In comparison, both segments have a greater Volume flexibility. However, this cannot be transferred to the whole system because of the common production dependencies that are necessary for maintaining the product mix. This leads to shortages in production, which are responsible for this situation and can be related to the workplace-levels. However, this will not be explained any further here, but will be part of the industry-related application of the evaluation methodology (see Sect. 4.2.3.1).

Table 3.14 Volume flexibility of the systems objects of the production-system in the example

System object	Factory					
F_{Volume} (Factory)	**27.44%**					
System object	*S1*			*S2*		
F_{Volume} (Segment)	36.42%			51.38%		
System Object	*L1.1*		–	*L2.1*		
F_{Volume} (Linie)	54.94%		–	51.38%		
System object	*WP1.1.1*	*WP1.1.2*	*WP1.0.1*	*WP2.1.1*	*WP2.1.2*	*WP2.1.3*
F_{Volume} (WP)	61.43%	32.93%	21.95%	67.15%	27.52%	29.22%

3.2.3 Flexibility Evaluation Method of Mix Flexibility

According to the approach of evaluation for Mix flexibility introduced in Sect. 3.1.2, its quantification has to be carried out with the help of the following parameters: the system optimal production profit, the profit optima which are limited by the production and finally with the help of the average production-profit- deviation. From this a production system's stability regarding the composition of the product mix can be determined. This provides information on the economical risk of endangering the success of the production by changing the mix.

3.2.3.1 Calculating the System-Optimal Production Profit

The purpose of calculating the system-optimal production profit is to determine the greatest profit that is achievable. It is used as a reference value for the subsequent determination of product-specific profit deviations. Consequently, an optimal production plan has to be found. This plan features, independently from a given product mix, the most profitable combination of products in terms of type and quantity that have to be manufactured. Such a production plan does not have to be calculated for each system object individually, but for the total system only, where it relates to the most profitable valid working hour model. The procedure for this is based on the basis algorithm (see Sect. 3.2.1) which, due to a changed general framework, is subject to small adaptations.

On the one hand, this concerns the objective function which has to guarantee that the total attainable profit of the production system is maximised. That is why it necessary to revert to the objective function presented in Formula 3.3 (p. 59) and this requires the determination of contribution margins c_i for each valid product-workplace -combination (see Formula 3.2, p. 59). The total contribution margin can be calculated using the optimal production plan, which provides the specific product-workplace-related output quantities x_i. As demonstrated in Formula 3.25, this margin becomes the system-optimal production profit after subtraction of the fixed costs.

Formula 3.25 Objective function for calculating the system-optimal production profit

$$c(x) = c^T \cdot x - C_{Fix} \rightarrow \text{Max.}$$

It is however not only the objective function that is influenced, but also the ratio conditions. These results from the limitations regarding a given product mix, that do not apply in this case. In contrast to the original causal connection that was part of the ratio matrix (see Formula 3.10, p. 64), the quantity ratio that is represented by the auxiliary variable t becomes meaningless. Only the component dependencies that are related to the following three conditions are important:

- A product has to be produced in that amount, that is required by the production of other products, because it is a component of them.
- An overproduction of intermediate products that cannot be sold is forbidden. They have to be produced in that amount only that, under the consideration of the scap, is needed as part of the manufacture.
- The production of end-products can be carried out in any random quantity.

Thus, according to Formula 3.8 (p. 63), two different kinds of conditions result in regard to the sellable quantity-ratios y_i:

- $y_i = 0$ holds for products that cannot be sold. This leads to them being sufficiently available as part of super-ordinate products.
- $y_i \geq 0$ is valid for products that can be sold. This is ensured by the production of these at any quantity, provided that they are at least produced in an amount which is required by the production of super-ordinate products.

In order to summarize the condition $y_i = 0$ of all non sellable products, a matrix $V_{ns} \in \mathbb{R}^{|IP| \times |R|}$ has to be defined whose single lines result from the scalar product of a vector y_i with production plan x. The matrix equation $V_{ns} \cdot x = 0$ is derived from this, and has to be transferred into a simplex-adequate notation (see also Formula 3.12, p. 66) according to Formula 3.26.

Formula 3.26 Ratio condition for non-sellable products for calculating the system-optimal production profit.

$$V_{ns} \cdot x \leq 0 \quad \text{and} \quad -V_{ns} \cdot x \leq 0$$

Correspondingly, a matrix $V_s \in \mathbb{R}^{|M| \times |R|}$ is also defined for products that can be sold. This enables the comprehensive illustration of the effective inequation conditions $y_i \geq 0$ by the matrix inequation $V_s \cdot x \leq 0$. However, since this inequation is not a simplex adequate representation due to the greater-than-or-equal-sign, it has to be transformed to a matrix inequation according to Formula 3.27.

Formula 3.27 Ratio condition for calculating the system-optimal production profit for sellable products

$$-V_s \cdot x \leq 0$$

Below the modifications that have to be carried out on the ratio conditions of the basis algorithm, in order to be able to determine the system optimal production profit, are exemplified with the help of the following example:

Example. According to the example production system in Sect. 6.4 products MI_1 and MI_5 cannot be sold individually. Their quantity ratios y_i can be calculated as follows:

- $y_{MI_1} = 0.99 \cdot x_{1,1} - 1 \cdot x_{2,1} - 1 \cdot x_{2,2} - 1 \cdot x_{3,2}$
- $y_{MI_5} = 0.98 \cdot x_{5,5} - 3 \cdot x_{6,6}$

When the matrix V_{ns} is arrayed, y_{MI_1} and y_{MI_5} are those lines in this matrix that have to be multiplied with the production-plan vector x:

$$V_{ns} \cdot x = \begin{pmatrix} 0.99 & -1 & -1 & -1 & 0 & 0 & 0 & 0 \\ 0 & 0 & 0 & 0 & 0 & 0 & 0.98 & -3 \end{pmatrix}$$
$$\cdot \left(x_{1,1}, x_{2,1}, x_{2,2}, x_{3,2}, x_{4,3}, x_{4,4}, x_{5,5}, x_{6,6}\right)^T$$

In order to fulfill the condition $y_i = 0$ which is valid for non sellable products, $V_{ns} \cdot x = 0$ also has to be valid. It is valid:

$$\begin{pmatrix} 0.99 & -1 & -1 & -1 & 0 & 0 & 0 & 0 \\ 0 & 0 & 0 & 0 & 0 & 0 & 0.98 & -3 \end{pmatrix}$$
$$\cdot \left(x_{1,1}, x_{2,1}, x_{2,2}, x_{3,2}, x_{4,3}, x_{4,4}, x_{5,5}, x_{6,6}\right)^T = 0$$

The same procedure is used for arraying matrix V_v for products that can be sold. They have to correspond to the condition $y_{MI_2}, y_{MI_3}, y_{MI_4}, y_{MI_6} \geq 0$. Thus, the matrix inequation $V_s \cdot x \geq 0$ that is valid for the example production system is:

$$\begin{pmatrix} 0 & 0.98 & 0.99 & -2 & 0 & 0 & 0 & 0 \\ 0 & 0 & 0 & 0.95 & 0 & 0 & 0 & 0 \\ 0 & 0 & 0 & 0 & 0.99 & 0.97 & -2 & 0 \\ 0 & 0 & 0 & 0 & 0 & 0 & 0 & 0.98 \end{pmatrix}$$
$$\cdot \left(x_{1,1}, x_{2,1}, x_{2,2}, x_{3,2}, x_{4,3}, x_{4,4}, x_{5,5}, x_{6,6}\right)^T \geq 0$$

Subsequent to the adaptations of the objective function that are in relation with the basis algorithm, the calculation procedure that is necessary to determine the system-optimal production profit can be expressed with the help of an optimization problem like in Fig. 3.9.

Requirements	Find the production plan $x \in \mathbb{R}^{|R|}$, which leads to a maximum objective function!
Objective Function	$c(x) = c^T \cdot x - C_{Fix}$
Constraints	▪ $\tilde{T}x \leq T_{max}$ ▪ $V_{ns} \cdot x \leq 0$ and $-V_{ns} \cdot x \leq 0$ ▪ $-V_s \cdot x \leq 0$ ▪ $x \geq 0$

Fig. 3.9 Simplex-adequate optimization problem for calculating the system-optimal production profit

Example. Applying the optimization problem from Fig. 3.9 to the example production system (see Sect. 6.4) provides a circular system-optimal production profit/gain $G_{opt} = 286{,}965.55$ MU, which corresponds to the production plan in Table 3.15:

Table 3.15 Production plan for a system-optimal production profit

System object	WP1.1.1		WP1.1.2		WP1.0.1	WP2.1.1	WP2.1.2	WP2.1.3
M_{WP}	MI_1	MI_2	MI_2	MI_3	MI_4	MI_4	MI_5	MI_6
$x(MI,WP)$ in QU	70,470	34,515	35,250	0	45,333	70,000	0	0
G_{opt}	**286,965.55 MU**							

3.2.3.2 Calculating the Product-Limited Profit Optimum

For each system object it must be specified what maximum profit a production system could still achieve if one of the products, that one can produce, is not produced. This has to be specified by means of the product-limited profit optimum. This requires for both the total system and for its subsystems, the formulation of a particular system-product-related optimization problem. Its solution requires a similar procedure to the calculation of the system-optimal production profit (see Sect. 3.2.3.1, which has been defined before. The aim is to maximize the achievable whole system profit with regard to all valid working hour models, without being bound to a given product mix. The only modification refers to addition of the extra constraints $x(MI, S) = 0$. It represents the situation where a chosen product MI within a certain system object S is not produced, so that its production quantity x takes the value 0. In order to convert this restriction to the simplex-adequate form of Formula 3.28, the transformation rules that have been mentioned in Formula 3.12 (p. 66) have to be applied.

Formula 3.28 Supplementary restriction for guaranteeing the not-production of a product within a system object

$$x(MI, S) \leq 0 \quad \text{and} - x(MI, S) \leq 0$$

By applying this additional constraint on the optimization problem in Formula 3.9 (p. 63), a new profit maximum is calculated for the whole system. Due to the additional restriction, the result can be less than before. This occurs when the not-production concerns an intermediate product that is essential for the production of another product. In summary Fig. 3.10 represents the simplex-adequate calculation model for determining the product-limited profit optimum with its valid constraints. The resulting value gets the term $G_l(MI, S)$.

Requirements	Find the production plan $x \in \mathbb{R}^{|R|}$, which leads to a maximum objective function!
Objective Function	$c(x) = c^T \cdot x - C_{Fix}$
Constraints	<ul><li>$\tilde{T}x \leq T_{max}$</li><li>$V_{ns} \cdot x \leq 0$ and $-V_{ns} \cdot x \leq 0$</li><li>$-V_s \cdot x \leq 0$</li><li>$x(MI, S) \leq 0$ and $-x(MI, S) \leq 0$</li><li>$x \geq 0$</li></ul>

Fig. 3.10 Simplex-adequate optimization problem for calculating the product-limited profit optimum

Table 3.16 Production plan of the product-limited profit optimum for product in MI_4 in segment S2

System object	WP1.1.1		WP1.1.2		WP1.0.1	WP2.1.1	WP2.1.2	WP2.1.3
M_{WP}	MI_1	MI_2	MI_2	MI_3	MI_4	MI_4	MI_5	MI_6
$x(MI, WP)$ in QU	70,470	34,515	35,250	0	45,333	0	0	0
$G_1(MI_4, S_2)$	**286,965.55 MU**							

Example. One example is to calculate the product-limited profit optimum for MI_4 from the point of view of $S2$ (see production system from Sect. 6.4). The result of this optimization problem from Fig. 3.10 is a maximum achievable profit of $G_1(MI_4,S_2) = 174,265.55$ MU. This corresponds to the following production plan in Table 3.16.

3.2.3.3 Calculation of the Mix Flexibility

The calculation of the index for the Mix flexibility for any system object in a production system is done on the basis of the mean production profit deviation. According to the calculation approach in Sect. 3.1.2 it is determined by the mean square deviation of all previously calculated product-limited profit optima. Care must be taken that only product-limited profit optima are part of the calculation, whose products can also be produced in the concerned system object. The following Formula 3.29 illustrates this.

Formula 3.29 Calculation of the average production profit deviation for a system object

$$\Delta G_a(S) = \sqrt{\frac{1}{n_M(S)} \cdot \sum_{MI \in S} \left(G_1(MI, S) - G_{opt}\right)^2}$$

whereas:

$n_M(S)$: Number of different kinds of products M of system object S;
$G_1(MI, S)$: Product-limited profit optimum of product MI in system object S;
$\Delta G_a(S)$: Average production profit deviation in system object S;
G_{opt}: System optimal production profit of the whole system

In order to guarantee the comparability of these average production profit deviations for different system objects or even for several production systems, they must be put in relation to the optimal profit according to Formula 3.30. Subsequently it has to be subtracted from value of 1, which stands for the complete Mix flexibility.

Formula 3.30 Calculation of the mix flexibility for a system object

$$F_{Mix}(S) = 1 - \frac{\Delta G_a(S)}{G_{opt}} \cdot 100\%$$

where:

$\Delta G_a(S)$: Average production profit deviation in system object S;
$F_{Mis}(S)$: Index of mix flexibility of system object S;
G_{opt}: System optimal production profit

At this point the area of validity of the application of Mix flexibility has to be emphasized. It has useful results when it is applied exclusively to a multi-product production. System objects with single-product fabrication get the value of 0.

Table 3.17 Mix flexibility of the system objects in the example production system

System object	Factory					
F_{Mix} (Factory)	**58.02%**					
System object	S1			S2		
F_{Mix} (Segment)	55.77%			77.33%		
System object	L1.1		–	L2.1		
F_{Mix} (Linie)	49.70%		–	77.33%		
System object	WP1.1.1	WP1.1.2	WP1.0.1	WP2.1.1	WP2.1.2	WP2.1.3
F_{Mix}(WP)	51.09%	86.29%	0%	0%	0%	0%

Example. Mix flexibility calculations for different system objects in the example production system in Sect. 6.4 result in the indices in Table 3.17.

According to these indices the whole system features a Mix flexibility of 58.02%. According to that, changes at the optimal production mix would lead to mean profit losses of 41.98%. Thus, the configuration of the product mix is not at all irrelevant for the efficiency of the system. This points to the existence of certain products within the fabrication spectrum that have a decisive influence on the production success. In this case, flexibility deficits are particularly found at workplace *WP1.1.1*. Their Mix flexibility mainly depends on product *MI₁*. Only this workplace can produce this product *MI₁*, which is part of other products that are sellable. In contrast workplace *WP1.1.2* is significantly less sensitive regarding the type of production and thus more flexible in regard to the mix. The consequences of such an evaluation are discussed in Sect. 4.2.3.3.

3.2.4 Flexibility Evaluation Methodology of Expansion Flexibility

Due to previous considerations in Sect. 3.1.3 the calculation procedure for quantifying the Expansion flexibility of production systems is orientated on the Volume flexibility. For this purpose, an expansion related flexibility range has to be determined that allows the economical expenditure of a constant increase of the output to be measured. It refers to the cost-benefit ratio of the best expansion alternative. Calculations that are necessary therefore concern the following parameters: The expansion related objective function, that represents the benefit aspect; and alternative specific break-even points, which represent different cost aspects.

3.2.4.1 Calculation of the Target Capacity

With the definition of a target capacity, a quantity-related expansion of the considered production system or of single subsystems thereof, is determined compared to the previous maximum capacity. Thus, expansion activities that do not achieve the target capacity can be excluded beforehand. Additionally, this also provides the opportunity to give different valid expansion alternatives a concrete value. This avoids the configuration of useless quantity-related flexibility potentials, because the break-even point is the exclusive criteria for the choice of an alternative (see Sect. 3.1.3).

Calculating the target capacity for system objects of a production system is carried out in the same way for all objects, by considering a given product mix. Under this consideration the evaluation method's user has to determine a percentile capacity expansion. This has to be multiplied with the previous maximum capacity that relates to this product mix, as Formula 3.31 demonstrates.

Formula 3.31 Calculating the target capacity for a system object

$$TC(S) = (1 + e) \cdot x_{MAX}(S)$$

whereas:

$TC(S)$: target capacity of system object S;

$x_{MAX}(S)$: maximum capacity for system object S;

e: Given capacity of expansion of system object S in per cent, specified by the user

   As mentioned before, those expansion flexibilities whose output is too low can be identified with the help of the target capacity. In order to do so, the same procedure as the one used for the calculation of the maximum capacity for Volume flexibility (see Sect. 3.2.2.1 is carried out and the maximum capacity is determined for each potential alternative. Planned expansions can either concern the whole system or a subsystem. Depending on the object concerned, an alternative specific optimization problem results according to Fig. 3.5 (p. 69) or rather Fig. 3.6 (p. 72). The simplex-algorithm is used for its solution. By comparing the calculated maximum capacity with the target capacity those expansion alternative that do not achieve the target value can be eliminated.

Example. To demonstrate the described context, segment *S2* of the example production system is used (see Sect. 6.4), whose previous maximum capacity has to be increased by 15%. For this purpose three expansion alternatives are specified in Sect. 6.4.3, and have to be checked for their goal fulfilment. They concern the set-up of a redundant production line (alternative 1), the set-up of an additional workplace (alternative 2) and the modification of workplaces (alternative 3).

   Calculating the maximum capacity for segment *S2* resulted in a maximum output of $x_{Max}(S2) = 116{,}267$ quantity units for product mix $MI_2 : MI_3 : MI_4 : MI_6 = 1 : 2 : 1.5 : 2$ that refers to the whole system. As a 15% capacity expansion is demanded for segment *S2*, $e = 0.15$. This leads to the following calculation:

$$TC(S_2) = (1 + 0.15) \cdot 116{,}267 \text{ QU}$$

Hence, the target capacity for segment *S2* is:

$$TC(S_2) = 133{,}707 \text{ QU}$$

   In order to check if those three potential expansion alternatives achieve the target capacity, the new maximum capacity, that segment *S2* could achieve with their implementation, has to be calculated for each of them. Accordingly, three separate, alternative optimization problems are formulated for system object *S2*, according to Fig. 3.6 (p. 72). Applying the simplex-algorithm leads to the following capacity-values:

- Alternative 1: $x_{MAX}(S_1) = 220{,}424$ QU
- Alternative 2: $x_{MAX}(S_1) = 133{,}843$ QU
- Alternative 3: $x_{MAX}(S_1) = 123{,}133$ QU

Comparing alternative-specific maximum capacities to the value of the target capacity shows that alternative 3 (modification of workplaces) leads to an insufficient output. That is why it will not be considered in further chapters.

3.2.4.2 Calculating the Alternative-Specific Break-Even Point

The purpose of calculating alternative-specific break-even points is to identify those alternatives among all valid expansion alternatives that relate to a specific system object, whose realization results in the best cost-benefit ratio. With the help of the break-even point, which is the decision criterion, it is shown which of the alternatives provides the least minimum production quantity for covering the total costs in the concerned production system (for a given product mix). This alternative then has to be used for calculating a concrete index for the Expansion flexibility of the considered system object.

Determining different break-even points $x_{BE_A}(S)$ uses the same method used for calculating the break-even point for the Volume flexibility. We must differentiate between a calculation procedure for the whole system and for subsystems. Thus, for each valid alternative the related simplex-adequate optimization problem is formulated in accordance with Fig. 3.5 (p. 69) or rather Fig. 3.6 (p. 72), and then solved with the simplex-algorithm afterwards.

Example. Returning to the example discussed last, the expansion alternative 3 of example production system in Sect. 6.4 has already been eliminated, because it did not achieve the target capacity. That is why only the break-even points of alternative 1 and alternative 2 have to be determined. By including the alternative-dependent parameters that are listed in Sect. 6.4.3, the following alternative-specific break-even points result for segment $S2$:

- Alternative 1: $x_{BE_1}(S_1) = 60,739\text{QU}$
- Alternative 2: $x_{BE_2}(S_1) = 57,383\text{QU}$

As a result, the expansion activity with the smallest break-even point would be the set-up of an additional workplace (alternative 3). Thus, with a 15% capacity expansion it has the best cost-benefit ratio for segment $S2$. That is why the expenditure that is associated with this expansion is the most economic.

3.2.4.3 Calculating the Expansion Flexibility

Finally, in order to be able to calculate the Expansion flexibility of a certain system object, the expansion-related flexibility range first has to be determined. After that, it has to be put in relation to the original flexibility range of the system object. While the dimension of the original flexibility range arises from the calculations for Volume flexibility (see Formula 3.23, p. 78), the expansion-related flexibility range is determined by quantifying the cost-benefit ratio of the best expansion activity.

For this purpose its break-even point has to be subtracted from the target capacity according to Formula 3.32.

Formula 3.32 Calculating the expansion-related flexibility range for different system objects

$$TC(S) - x_{BE_A}(S)$$

Formula 3.33 is used to determine the index of the Expansion flexibility of the considered system object with the help of the determined expansion-related flexibility range. This also guarantees its comparability.

Formula 3.33 Calculating the Expansion flexibility for different system objects

$$F_{Expansion}(S) = \frac{TC(S) - x_{BE_A}(S)}{x_{MAX}(S) - x_{BE}(S)} \cdot 100\%$$

Example. Based on the two previous examples for determining the target capacity and alternative-specific break-even points, it should be aimed for the set-up of an additional workplace (alternative 2) for the expansion of the maximum capacity in segment $S2$ in the example production system (see Sect. 6.4). Under the consideration of the given product mix for the whole system, which is $MI_2 : MI_3 : MI_4 : MI_6 = 1 : 2 : 1.5 : 2$, the following expansion-related flexibility range results:

$$TC(S_2) - x_{BE_A}(S_2) = 133,707 \text{ QU} - 57,383 \text{ QU} = 76,324 \text{ QU}$$

As the dimension of the original flexibility range with:

$$x_{MAX}(S_2) - x_{BE}(S_2) = 116,267 \text{ QU} - 56,532 \text{ QU} = 59,735 \text{ QU}$$

Knowledge from previous Volume flexibility calculations shows that the following index for the Expansion flexibility in segment $S2$ results:

$$F_{Expansion}(S_2) = \frac{76,324}{59,735} \cdot 100\% = 127.77\%$$

Due to the fact that the Expansion flexibility in segment $S2$ is greater than 100%, the conclusion is that the intended expansion in regard to the segment leads to an advancement of the Volume flexibility. Following the distinction of cases concerning the change of dimension of their expansion-related flexibility range as described in Sect. 3.1.3, the system object is considered to be "completely expansion-flexible". The effect on the flexibility and on other system objects that results from these expansions are discussed in Sect. 4.2.3.2. Furthermore, possible applications and utility of the indices that can be calculated with this evaluation procedure will be explained in detail.

3.3 Definition of the Cost Accounting Reference Frame

According to the above-mentioned thoughts on evaluation methodologies of Volume-, Mix-, and Expansion flexibility, determining system-dependent flexibility indices traces back to specific optimization problems. They can be solved with the help of the simplex-algorithm. Such an optimization problem is a model that is established mathematically and that contains logical relationships and that is described by parameters, which refer to those characteristics of production systems which describe flexibility. As variable elements, to which certain data are allocated, these parameters are structured in three basic groups, as explained in Sect. 3.2.1.1: non-cost related, cost related and user-dependent calculation parameters (see Table 3.2, Table 3.3, Table 3.4 on pp. 57–58). Since determining their concrete values directly influences the quality of the results of flexibility calculations, this is executed according to defined criteria. Above all, this concerns cost related calculation parameters, whose value assignment is especially dependent on the amount of the product related technological value added and revenues. In order to be able to calculate them as exactly as possible in the context of flexibility analysis, a previous gathering and structuring (apportionment) of variable and fixed cost elements in production systems, that are connected with the value added, is required. Here the cost accounting that is established in business economics is especially suitable. A main characteristic of cost accounting is the consideration of all economically evaluable operations of realized or planned business processes, for which purpose a budget is established. This budget is established on the basis of a special scheme that is also the database of different business analyses (vgl. [Eber-04] [Götz-04]).

Similarly, in order to guarantee a standardized, complete and transparent cost structuring for the evaluation methodology, which leads to the calculation of reproducible Volume-, Mix-, and Expansion flexibility, such a chart of accounts is also required. From hereforth it will be called the *cost accounting reference frame*. This frame allocates different cost elements, which occur in a production system, to cost-related calculation parameters (see Table 3.3, p. 57) of single evaluation-relevant objects.

3.3.1 Choice of a Cost Accounting Procedure

Different cost accounting procedures exist in theory and practice that enable the setup of the intended cost accounting reference frame. That is the reason why it has to be clarified which of the potential procedures is best suited for the intended interests of cost structuring within the evaluation methodology. You can find a more detailed description of basic procedures of cost accounting in Sect. 6.1.1.

3.3.1.1 Decision Between Full Cost and Direct Cost Accounting

Basically, cost accounting procedures can be divided into full cost accounting and direct cost accounting depending on their degree of attribution (see Sect. 6.1.1.1). Both main types of cost accounting are suitable for cost accounting reference frame in principle, especially as they are both applied in practice.

Cost accounting mainly originated from *full cost accounting* whose purpose is to maintain a financial equilibrium in companies. Executing the three calculation steps cost type-, cost centre-, and cost object accounting (see Sect. 6.1.1.2) leads to an overview in regard to type, amount and structure of the internal use of resources, as well as the places where costs and benefits originate. The advantage of this is an appropriate apportion of costs for calculating flexibility indices. However, the fair apportion between originators of the operating costs to respective cost centres is very difficult, as inaccuracies and failures may occur. From this the risk of imprecise results of flexibility calculations results. In addition to that, the missing considerations of the utilisation when operating costs are distributed to cost objects, have to be considered critically. This is not problematic when it comes to variable costs. However, problems might occur with fixed costs, as in the context of flexibility evaluations the fixed costs attribution per piece depends on the utilisation. The most serious drawback is the fact that full cost accounting is not suitable as a basis for enterprise decisions and that it only considers market data insufficient (see [Eber-04] [Hofm-04]. However, this is the request of evaluation methodologies that are to be designed. In its role as a support in decision making in production systems, it has to provide reliable estimations about flexibility in production systems, so that reaction to market uncertainties can be carried out on time.

In comparison, *direct cost accounting* is considered to be more focused on competition and allows a comparatively easy apportion of arising expenses in variable and accordingly fixed elements (see Sect. 6.1.1.3). This allows an easy and fast allocation of evaluation relevant objects in a production system to the cost-related calculation parameters. In contrast, applying full cost accounting requires the additional setup of a cost function that is as close to practice as possible and which has to be connected with different optimization problems that are presented in Sect. 3.2 However, since finding such a cost function alone involves high uncertainties, one can assume that the results of direct cost accounting are more exact and less faulty. Another advantage of direct cost accounting is a better consideration of change in utilisation, as well as a more simple involvement of scenario-based additional costs, for example costs due to expansion activities. However, direct cost accounting features certain disadvantages too, and is thus considered to be inappropriate or applicable conditionally only for certain actions, for example the long-term price determination or the determination of profit for pieces and orders (see also [Götz-04] [NN-05]). In the context of this book, these deficits are of no consequences and that is why they are chosen.

Due to the existing subdivision of direct cost accounting into the three basic calculation systems: direct cost accounting with global consideration of fixed costs, direct cost accounting with differentiated considerations of fixed costs and relative direct cost accounting (see Fig. 6.2, p. 153), there is the need for the additional identification of the most suitable of these cost accounting systems. Limited by the request of the evaluation methodology that a flexibility evaluation should be carried out from workplace level to factory level (see Sect. 2.5.2), different organisational hierarchies in a production system have to be considered. That is why a great number of cost types arise, whose complexity increases due to the consideration of a multi-product-production, which also complicates a sensible apportion. That is why the three systems of direct cost accounting are evaluated by their suitability for gathering and structuring costs on different consideration levels.

3.3.1.2 Choice of a Direct Cost Accounting Procedure

Direct cost accounting with global considerations of fixed costs (see Sect. 6.1.1.3.1) only roughly classifies costs and differentiates between product-dependent variable costs and global fixed costs. The latter are allocated to all products, which leads to the fact that a product- and product group dependent allocation of fixed costs, as required in a multi-product-production, is not possible. Furthermore, due to the block of fixed costs that cannot be apportioned, no level-specific allocation of costs to workplaces, lines, segments and the factory can be carried out. That is why the information value of this procedure is not sufficient to meet the required task.

Compared to this, direct cost accounting with differentiated considerations of fixed costs (see Sect. 6.1.1.3.2) and the relative direct cost accounting (see Sect. 6.1.1.3.3) allow the apportion of fixed costs while considering product- and product-group-related dependences, as well as their allocation to different organizational hierarchies. Thus, both calculation procedures are recommended for an allocation of several consideration levels. However, direct cost accounting with differentiated consideration of fixed costs is preferred, as it is widely applied in practice. The reason for the lack of acceptance of the Riebelsche calculation system is that the differentiation of the fixed costs block is too high. Oftentimes this leads to insurmountable problems in practical work, which complicates the establishment of the periodical calculation of earnings before taxes and interests. However, as the practical suitability is a main criterion for the success of the evaluation methodology, direct cost accounting with differentiated consideration of fixed costs has to be applied when the cost accounting reference frame is defined.

3.3.2 Determining the Hierarchies of Reference

According to procedure of direct cost accounting with differentiated considerations of fixed costs, the variable and fixed parts of costs that have to be structured need to be classified in objects of attribution that establish a hierarchy of objects (see Sect. 6.1.1.3.2). Determining this hierarchy is based on the requests of the evaluation methodology, which requires that flexibility calculations must be analysed level-specifically (see Sect. 2.5.2). Consequently, workplaces are considered as the lowest

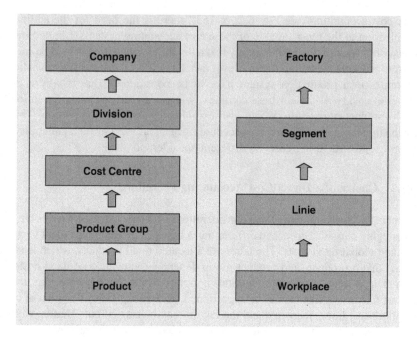

Fig. 3.11 Comparison of reference hierarchies of the chosen direct cost accounting procedures (on the *left*) and the cost accounting reference frame (on the *left*)

cost accounting object, whose superior level of cost accounting is the object line, which can include several workplaces. According to the consideration levels of the production systems that are described in Sect. 2.1.3, the object segment is the next level above the object line. Different lines or single workplaces without connection to lines can belong to the segment, so that production-dependent cost elements that have not yet been calculated on a workplace- or line level are redistributed. The highest level of the hierarchy is the factory object, to which all remaining costs inevitably have to be allocated.

Fig. 3.11 summarizes single levels of the hierarchy of cost objects of the cost accounting reference frame and compares them to the proposed reference levels from direct cost accounting with differentiated considerations of fixed costs.

3.3.3 Definition of Basic Cost Types

In accordance with the classification of reference hierarchies, cost types can be determined which summarize the main information of costs in each level in a practical way. However, only those costs that are connected either directly or indirectly with the added value of the production system are considered. Provided that they are not part of the production, as for example financial investment or product-independent services (e.g., reparation-/maintenance orders of similar products), their costs and benefits remain unconsidered. The same applies for the production development and construction, which are not focused on in this book.

The choice and structuring of these types of costs is carried out on the basis of production systems resources that are identified in Sect. 2.1.2 and on the basis of the executed analyses in practice, in the form deemed useful by the author. Here, a high level of abstraction was chosen to ensure that a universally valid and company-independent apportion of costs is possible and which, if needed be, can also be detailed company-specifically.

The following lists all the cost types that are considered to be relevant and that are structured according to the chosen reference hierarchies.

3.3.3.1 Workplace Level

As already demonstrated in Sect. 2.1.3, only the actual value added i.e., the physical value added, is performed here. Related cost types that have to be attributed to the cost accounting reference frame concern:

1. *Product-related material costs on the workplace level*

A main aspect that has to be considered when a manufactured item is produced, is its material costs. According to the term material these are raw materials, auxiliary materials, operating supply items or components/construction groups (see Sect. 2.1.2.3). That is why material costs that are related to a product include the direct costs of the required material type. Operating supply items need to be considered separately since, as opposed to other material types, they are not part of a product and except for benefit-related energy costs (energy carrier like fuel or electric current), they can hardly be allocated to a product. That is why they are treated separately and entered somewhere else, except for energy costs[6] that can be attributed directly.

Since producing a product can often be carried out at different workplaces, the amount of material costs for the same product can differ due to production-specific conditions. That is why cost accounting needs to be guaranteed separately in dependence on the product and the particular workplace. Formula 3.34 demonstrates how material costs for a so-called product-workplace-combination are calculated.

Formula 3.34 Calculating of product-related material costs per workplace

$$C_{Mat}(MI, WP) = C_{raw} + C_{aux} + C_{supply} + C_{comp} + C_{Energy},$$

whereas:

$C_{Mat}(MI, WP)$: Material costs for a product-workplace-combination;
C_{raw}: Raw material costs for a product-workplace-combination;
C_{aux}: Auxiliary material costs for a product-workplace-combination;

---

[6]The procedure of calculating product-related energy costs is explained in Sect. 6.3.1.

C_{supply}: Operating supply items cost for a product-workplace-combination;

C_{comp}: Components/construction groups costs for a product-workplace-combination;

C_{Energy}: Energy costs for a product-workplace-combination

2. Product-related labour costs on the workplace-level

This type of costs refers to wages of those employees who are involved in the physical production via their workplaces, as for example operating- or setup labour. Normally, these labourers get working hour related wages instead of a salary package (see Sect. 2.1.2.2). Thus, product-related labour costs can vary, depending on the time of day and working hours. In order to capture these costs in terms of the workplace, Formula 3.35 has to be applied. According to this, the costs of wages per time unit of an employee of a considered product-workplace-combination have to be multiplied by the product-related cycle time (see Sect. 6.3.2) at the workplace. Furthermore, a factor for additional labour costs (see Formula 3.36) has to be considered so that working time-conditioned wages are also taken into account, for example shift work, as well as a time-related factor of employment.

Formula 3.35 Calculating of product-related labour costs

$$C_{labour}(MI, WP) = \sum_{W} \left(c_{wage/t}(W, WP) \times t_{CT}(MI, WP) \times aLC \times B \right)$$

whereas:

$C_{labour}(MI,WP)$: Labour cost for a product-workplace-combination;

$C_{wage/t}(W,WP)$: Wage for a worker per time unit for a workplace;

t_{CT}: Cycle time for a product-workplace-combination;

aLC: Factor for additional labour costs for a worker-workplace-combination;

E: Time-related factor of employment for a worker-workplace-combination

The time-related factor of employment is based on the fact that especially automated workplaces do not necessarily feature the same number of persons during their operating time. If one employee is only 50% involved in the production at a certain workplace, the factor of employment of 50% will be allocated to him. In order to finally obtain the total labour costs for a product-workplace-combination, labour costs of all the employees working there have to be added up (see Formula 3.35).

Formula 3.36 Calculation of the factor for additional labour costs

$$aLC = \frac{t_r \times C_{r\_wage}(W, WP) + t_a \times K_{a\_wage}(W, WP)}{(t_r + t_a) \times K_{r\_wage}(W, WP)},$$

whereas:

$aLC_{W,WP}$: Factor for additional labour costs;

t_r: Regular working time;

t_a: working time with additional labour costs;
$C_{r\_wage}(W,WP)$: Costs for a worker without additional labour costs for a workplace;
$C_{a\_wage}(W,WP)$: Costs for a worker with additional labour costs for a workplace

3. *Product-independent material costs on the workplace-level*

These are the material costs at a workplace, whose amount changes depending on the volume of the production, but whose detailed apportion is very complicated depending on the kind of product. Above all, this concerns operating supply items like coolants, lubricants, cleaning agents, etc. (see Sect. 2.1.2.3). This also comprises costs for workplace-related office supplies like paper, pens or books and also energy costs that cannot be related to an explicit product-workplace-combination.

4. *Production equipment costs on workplace level*

Depending on the kind and usage of a workplace it can either be considered as independent production equipment, or it can combine several pieces of production equipment. Capturing its costs within the cost accounting reference frame is done in the form of depreciation. As already explained in Sect. 2.1.2.1, production equipment can be classified in two basic groups according to their participation in the actual added value. These two groups are production equipment that is directly and indirectly involved in the production. Both can be considered at a workplace level and will have the same outcome, independently from the kind- and amount of the product that is produced by them.

(a) Costs production equipment that used directly in the production

Costs that result from this refer to workplace specific means of production like machines, devices and tools that contribute directly to the production in form of a physically value added.

(b) Costs production equipment that are used indirectly in the production

These costs result from objects that, in contrast to means of production, do not perform any kind of physical machining operations on the products. Instead they contribute to the product-related organization of a workplace and can in turn be attributed to it. Examples are factory equipment like mobile file pedestal, chairs or other utilities and cleaning equipment too.

5. *Maintenance costs of the production equipment on the workplace level*

In order to maintain the functional status of the workplaces or in order to be able to re-establish it in case of a breakdown, the so-called maintenance activities are needed. They do not only comprise a periodical attendance and inspection, but also the corrective maintenance in case of disturbances or damages. The latter does not have to refer to the removal of technical damages only, but can include repair tasks for production equipment. Resulting maintenance costs of a workplace are to be considered as product-independent. Due to varying operating times, their amount can differ in single accounting periods.

6. *Expansion costs on the workplace level*

In order to provide the most exact capturing of cost-effective expansion activities as possible with regard to the calculation of Expansion flexibility, a separate type of costs is needed within the cost accounting reference frame. All these cost elements have to be directly connected to technical or organizational modifications, which constantly increase the output. As a rule, they arise in the context of projects that are especially initiated for this purpose, that develop and realize measures that effectuate a targeted benefit improvement. This does not include costs that, due to their character, belong to other types of costs. For example this concerns acquisition costs of an additional machine module for increasing the variety of products. They are considered as "costs of production equipment that are directly involved in the production" since they are subject to special fiscal depreciation.

Expansion costs relate much more to additional expenses that are based on planning and realizing corresponding expansions. Amongst others, this includes costs for engineering (in-house expenditure and/or external service), the coordination and organization of expansion activities and also start-up costs.[7] Provided that a workplace is expanded organizationally as for example when human resources are increased, additional costs for on-the-job-training, courses of instruction or one-off costs for the acquisition- and organization of the labour apply. As such costs have a planning background that is confronted with a use of long duration, they have to be allocated equally for the planned action time of an expansion activity (see Fig. 2.7, p. 26). This is similar to the depreciation of production equipment (see Sect. 2.1.2.1).

3.3.3.2   Line Level

This hierarchy that is characterized by a standardized linking of workplaces in the interest of a linked production flow (see Sect. 2.1.3.2) considers the following types of costs as fundamental:

1. *Product-independent material costs on the line level*

All materials that can be allocated to the line level are product-independent material costs. They are considered to be costs for maintaining the line-specific readiness for action and for their function execution. That is why apportioning them in a product-workplace-related way is not practical, even if they change depending on the produced kind of product and on the amount of production. This includes

---

[7]Start-up costs result from additional costs that are needed for the production during the starting period or when new/modified machines are brought into service compared to productions where none of these situations apply. They result from the higher scrap and a longer productive time per quantity unit [Adam-98].

costs for operating supply items as for example the usage of line-specific transport system/means of transport or for heating/cooling and illuminating the environment of the production or premises. Furthermore office equipment is included, such as paper, pens, books and other material that ensures the readiness of action/the operation execution of the line in any way.

2. *Product-independent labour costs on the line level*

Aside from the workplace labour within a line, further labour is generally needed that takes care of planning, organization and control of the line and for leading the workplace employees. In contrast to the workplace related staff, the costs caused by this are fixed costs instead of variable costs and have to be allocated to the product-independent labour costs of the line. The reason is that the distribution of costs to single products can be only be executed in a limited way. However, the amount of costs can vary depending on the working hour model used (see Sect. 2.1.2.2). For example, the wage of a shift supervisor of a production line corresponds to the working hours and the time of day.

3. *Costs of production equipment on the line level*

Gathering costs of production equipment at the line level is executed in the same way as at the workplace, i.e., on the basis of depreciation. That is why they do not depend on kind- and amount of the production. However, it has to be considered that production equipment at the line level only has an indirect relation to the physical value added and that in this cost type, only production equipment costs that are indirectly involved in the production are gathered. These include depreciation of transport systems and transport utilities for a continuous or discontinuous material flow of the line. Examples are band-conveyors or vehicle systems, but also iron-barred boxes and pallets. Furthermore, depreciation of factory equipment and other production equipment are part of it, because they could possibly arise at the workplace level.

4. *Maintenance costs of production equipment on the line level*

Analogous to workplaces, maintenance costs also arise at the line level. These mainly refer to existing transport systems/means of transport. Their purpose is to maintain the availability for use or to reconstruct it in case of operational deficiency. Depending on the intensity of their use, their amount can differ.

5. *Expansion costs on the line level*

Specific technical and organizational modifications whose purpose is to increase the output are not only limited to the workplace level. They can also be based on a line, due, for example, to the establishment of a new workplace that expands an existing linking of lines. Similarly to the workplace level, this also requires adequate planning and organization for executing these kinds of expansion activities. Resulting financial expenses are calculated equally for the duration of the action time, except for those that have to be allocated to other cost types.

3.3.3.3 Segment Level

According to the underlying understanding of production segments in this book, the different production units (lines and workplaces) are summarised due to their product-orientated independence (see Sect. 2.1.3.3). In a sense they are similar to lines (linking of lines) since their cost types of this hierarchy are identical to those of the line level:

1. *Product-independent material costs on the segment level*

   Material costs that have to be allocated to the segment level are characterized by their fundamental function of maintaining the operational readiness and the operation execution of a segment. That is why they are exclusively attributed to it, and exclude gathering material costs that are already calculated on other levels. This includes costs for operating supply items (for example segment specific transport systems/means of transport or the illumination of the production- and administration area), office- and other material costs, that are considered to be product-independent for the same reasons as on the line level.

2. *Product-independent labour costs on the segment level*

   Labour costs on the segment level are basically connected to a segment-related, indirect operation that is close to the production. In principle identical cost elements as on the line level are entered, while additional cost aspects are also considered, for example: administration costs of a segment specific store. Furthermore, one must assume that labour costs that can be gathered on the segment level are composed rather of salary packages than of (efficiency-oriented) wages (see Sect. 2.1.2.2). The reason is that there does not have to be a direct connection between the processing in terms of time of segment-related duties and the moment of the actual production. Thus, the salary of a works director or its assistance can be independent from the working hours and the moment of work in contrast to the wage of a shift supervisor of a line. Accordingly, a detailed apportion of costs is rather difficult for this type of costs, which is why segment-related labour costs are considered to be product-independent. Nevertheless their amount can vary depending on the working hour model used.

3. *Production equipment costs on the segment level*

   Production equipment that is used in the segment and which is not directly involved in the technological value added has to be considered as production equipment costs that are indirectly involved in the production. The calculation of their costs makes use of the same method as at the lines, by means of depreciation, independent of the type and amount of the produced products. Examples are transport systems/means of transport, that are used only for the segment-specific linking of the production flow and that cannot be added to a single line due to that reason.

4. *Maintenance costs of production equipment on the segment level*

   Maintenance costs can accrue for production equipment on the segment level too. Calculation of the resulting costs is done in a similar way to the costs on the workplace or line level. They are considered to be product-independent, as it is not impossible that they might vary in regard to their frequency of occurrence or their amount of costs.

5. *Expansion costs on the segment level*

Engineering-, organization-, starting and other costs that are linked to a segment-related expansion activity have to be allocated to this type of costs. An example could be the build-up of an additional production line in order to increase the output of one or several products. As with the expansions on other levels, costs that are associated with this are distributed equally over the duration of the expansion-dependent action time due to their planning character.

3.3.3.4 Factory Level

Costs that arise here have their origin in the indirect, central processes of the production system, according to the knowledge of Sect. 2.1.3.4. They only contribute indirectly to the production system, but are still inseparably linked with the production. That is why costs that are caused by them still need to be gathered. As mentioned at the beginning of the chapter, implicit production costs that are not relevant for the production, but which might still arise in the context of the factory, remain excluded. These costs must not be considered in this cost allocation. The following cost types are valid for this hierarchy:

1. *Product-independent material costs on the factory level*

In addition to the material costs that have to be added to the other production hierarchies, there is also a material consumption on this factory level, that has to be gathered separately. The material consumption on this level results less from direct production activities or operations within the production system, but much more from central tasks of the factory organization like accounting, human resource management, factory logistics, central store etc. (see Sect. 2.1.3.4). At the same time office- and other materials are needed in a considerable amount, as well as operating supply items, that are needed for illuminating the factory site and local buildings or for heating, cooling and air ventilation. Due to the fact that all factory-related material costs can be apportioned in a limited way only, these costs are considered to be product-independent, however they might vary in dependence on the production volume.

2. *Product-independent labour costs on the factory level*

In principle these are labour costs that result from personnel that is involved in the production system and that do not belong to any of the previous production levels. They are considered to be product-independent and are specifically connected to the factory level. Although this kind of labour costs, in contrast to those of the lines and segment are made up mainly of salary, work-related wages that are directly connected to this level cannot be excluded. Thus, it is improbable that a part of the workforce receives wages from the central stock, whose amount might vary in dependence on the used working hour model.

3. *Production equipment costs on the factory level*

'Due to the fact that production equipment might have to be available for the tasks on the factory level if needed, the costs caused by them have to be gathered as a separate cost type. However, this is based on the assumption that they are only indirectly involved in the production, as is the case on the line and segment level, because the physical value added takes place on workplace level. The total amount of the costs that result from this is determined, as explained with the other production levels, with the production equipment related depreciation which is considered as product-independent as explained before.

4. *Maintenance costs of the production equipment on the factory level*

Maintenance measures that need to be executed on the factory level are directly related to the production equipment that is gathered here. This includes attendance- and service jobs of factory-related transport systems and of means of transport or of building- and property equipment. Similar to the previous production levels, resulting costs are subject to fluctuations with regard to their amount and are product-independent.

5. *Expansion costs on the factory level*

Costs for expansion activities that do not specifically refer to a workplace, a line or a segment, but concern the production system as a whole and cannot be allocated to another cost type due to their character, are summarized in this cost type. This includes engineering-, organization-, and start-up costs, which result from a production-related and an organizational reconfiguration of the production system, for example the installation of a new production segment. As these kinds of costs cause a long duration of action time, they are allocated equally over the planned period of use.

6. *Environmental costs of the production systems*

The Environmental costs are closely related to the system environment of the production. They do not only trace back to the lawmaker, but in light of the environmental awareness that has strongly changed in the past years, have an important market economical background. Therefore the importance of environmental costs of a production system, which are gathered in a separate cost type, should not be underestimated. This includes costs for environmental taxes and licenses, surveillance and dealing with emissions, product-related disposal management and other environmental protection activities. They concern the production system as a whole only and cannot necessarily be allocated to single workplaces, lines or segments. That is why the consideration of all system-related environmental costs is carried out on the factory level. In doing so, it is assumed that they are product-independent, because their amount might vary due to varying operating times in the different accounting periods, similar to the maintenance costs.

3.3.3.5 Graphical Summary of the Cost Types

Table 3.18 summarizes the classification of all afore-mentioned cost types of a reference level in the cost accounting reference frame.

3.3.4 Classification of Defined Cost Types in Cost Categories

By defining the cost accounting reference frame for a production system it is determined which cost elements have to be allocated to its system objects that are organized on different hierarchy levels. However, in order to determine concrete values for cost-related calculation parameters (see Table 3.3, p. 57) for such a system object, it is necessary to categorize the cost types in variable and fixed costs. The procedure for determining the variable and fix parameter values, which is a fundamental condition for applying the evaluation methodology for Volume-, Mix-, and Expansion flexibility (see Sect. 3.2), shall be the subject of the following explanations.

3.3.4.1 Determining Variable Costs

A basic characteristic of variable cost elements that are their *ability to be definitely allocated to a product and workplace*. Accordingly, their contribution changes depending on the type and amount of the production. From the point of view of the cost types that were defined in Sect. 3.3.3 this concerns the product-related material costs as well as the product-related labour costs. Both exclusively occur for system objects on the workplace level, and increase continuously with each produced product. Thus, their variable total costs can be calculated from the sum of the material- and labour costs of a product. Since their amount can depend on the workplace that produced the product, the particular product-workplace combination has to be considered (see Formula 3.37).

Formula 3.37 Calculation of variable costs of a product-workplace-combination

$$C_{var}(MI, WP) = C_{Mat}(MI, WP) + C_{labour}(MI, WP),$$

whereas:

$C_{Mat}(MI,WP)$: Material costs for a product-workplace-combination;
$C_{labour}(MI,WP)$: Labour costs for a product-workplace-combination;
$C_{var}(MI,WP)$: Variable costs for a product-workplace-combination

A problem in applying Formula 3.37 is that it does not include different working hour models that effect the variable costs of a product-workplace-combination which deviate from each other, as is demonstrated by the following example:

Table 3.18 Level-related classification of basic cost types within the cost accounting reference frame

Cost categories and cost types	Cost objects (references hierarchies)	Workplace level			Line level			Segment level			Factory level
		WP 1	WP 2	WP X	Line 1	Line 2	Line X	Segment 1	Segment 2	Segment X	Name of factory
Variable costs	Product-related material costs										
	Raw materials										
	Auxillary materials										
	Supplies										
	Components parts/componeny groups										
	Enregy										
	Product relaed labour costs										
"No real" fixed costs	Product-related material costs										
	Labour costs										
	Maintanence costs of production equipment										
	Environmental costs										
"Real" fixed costs	Cost of production equipment										
	Directly involved in production										
	Indirectly involved in production										
	Expansion costs										

▓ Heirarchy of reference considers the cost type; ☐ Heirarchy of reference does not consider the cost type

Example. A production system runs a three shift system (24 h per day) from Monday till Friday, which causes the total variable costs of 2 MU for a certain product-workplace-combination. This already includes a wage increase for night work. Should this working hour model change, in order to achieve for example a maximum production equipment utilisation with a four/or five shift system, it could increase the variable costs of the considered product to $2 + x$ MU, as it causes additional wage increases for working on the weekend.

As these fluctuations can decisively influence the results of flexibility calculations, variable costs have to be considered separately for each working hour model. Accordingly, Formula 3.37 is expanded by a working time index *WHM* according to Formula 3.38:

Formula 3.38 Calculation of variable costs of a product-workplace-combination in dependence on the working hour model

$$C_{var, WHM}(MI, WP) = C_{Mat, WHM}(MI, WP) + K_{labour, WHM}(MI, WP)$$

Another important aspect that influences the amount of variable costs of a workplace-product-combination is their specific scrap rate t_{max}. The amount of the scrap rate can also vary depending of the used working hour model. As the scrap rate of a workplace can be budgeted both for its production costs as well for its production quantity, it is not important when the costs are determined. Instead, it is considered as part of the non-cost related calculation parameters (see Table 3.2, p. 57) and thus influences the results of the flexibility evaluations.

3.3.4.2 Determining Fixed Costs

After the previous classification of variable costs in product-related material and product-related labour costs, all remaining cost types can, in principle, be allocated to the category of fixed costs. However, this requires the previous differentiation of the two different groups of fixed costs. The reason is that cost elements in a production system exist that cannot be clearly apportioned to a product-workplace-combination. However, their amount still changes depending on the type and amount of the production. That is why such costs shall be allocated to the group of *"non-real" fixed costs*. Examples could be varying costs for operating supply items or maintenance costs when the working hour model is changed. Consequently, the other group of fixed costs comprises the so-called *"real" fixed costs* that remain constant, independent of changes in the production amount or composition of the product mix. A typical example is costs of depreciation of production equipment.

Table 3.19 shall give an overview of the categories of the cost types "real" and "non-real" fixed costs, as well as variable costs that were defined in Sect. 3.3.3.

Table 3.19 Classification of cost types that are defined in the cost accounting reference frame in "real" and "non-real" fixed costs

Variable costs	"Non-real" fixed costs
Product-related material costs on the workplace level	Product-independent material costs on the workplace-level
Product-related labour costs on the workplace-level	Maintenance costs of the production equipment on the workplace level
"Real" fixed costs	Product-independent material costs on the line level
Production equipment costs on the workplace level	Product-independent labour costs on the line level
Expansion costs on the workplace level	Maintenance costs of production equipment on the line level
Costs of production equipment on the line level	Product-independent material costs on the segment level
Expansion costs on the line level	Product-independent labour costs on the segment level
Production equipment costs on the segment level	Maintenance costs of production equipment on the segment level
Expansion costs on the segment level	Product-independent material costs on the factory level
Production equipment costs on the factory level	Product-independent labour costs on the factory level
Expansion costs on the factory level	Maintenance costs of the production equipment on the factory level
	Environmental costs of the production systems

In order to be able to use both the "real" and the "*non-real*" fixed costs of a system object as a (fix-) cost related calculation parameter (see Table 3.3, p. 57), they have to be standardised to a pre-defined analysis period (e.g., week or quarter). This period must demonstrate a representative mean of the total period under review. For this purpose it is recommended to determine all fixed costs of a system object for a year of operation, and then divide them by the number of total periods. For each working hour model this needs to be executed separately, as represented by Formula 3.39.

Formula 3.39 Determining level-related (standardised) fixed costs

$$C_{Fix,WHM}(S) = \frac{\sum\limits_{1\ year} K_{Fix,WHM}(S)}{n_P},$$

whereas:

S: System object (e.g., workplace or segment)
P: Period
n_P: Number of periods within a year
$K_{Fix,AZM}(S)$: Standardised fixed costs per period for the system object S in working hour model *WHM*

$\sum\limits_{1\ year} C_{Fix,WHM}(S)$: *Annualised Sum of fixed costs ("real" and "non-real") for he*
system object S in working hour model WHM

3.4   Concept of the Production System Model

A topic open to further discussion is the conception of a production system model, hereafter referred to as PSM, to obtain a data connection between the real-world analysis object and the three flexibility evaluation methods described in Sect. 3.2. According to the associated requirements (see Sect. 2.5) the specific relationships and dependencies of the various system objects are to be mapped. The provision of all necessary information on the object-dependant flexibility calculations at different levels of observation must be ensured in a form that allows the fast and flexible flexibility measurements of different types of production systems, including their subsystems. In this context, the abundance of potential products and the diverse organisational forms of the system object set high demands on the free configuration ability of the PSM. To meet the challenges that this poses, the conceptual basis of the model forms a so-called object-orientated reference model which allows, using the hereditary principle, the dynamic configuration of the PSM at the respective structure of the production system being evaluated.

3.4.1   The Object-Orientated Reference Model

The initial idea for the object-orientated reference model is based on a basic hierarchical classification of production systems, as shown in Sect. 2.1.3 in the breakdown into factory, segment, line, and the workplace. However, as a result of the various system structures, deviations from this have to be allowed. Under the current understanding, the system object "factory" always represents the entire system. A workplace however, is not necessarily a component of a line. It could just as easily have a direct affiliation with a segment or a factory, without any line reference. In order to represent such mapping relations in the PSM, the paradigm of object orientation for the reference model can be used. This is described in more detail in the Sect. 6.1.3. A key advantage of this method lies in the relatively large, industry-independent freedom of movement in the construction and parameterization of the various system objects. Thus the flexibility ratings for both the model-specific adaptations and adjustments of the flexibility metrics can be easily executed.

The general structure of the reference model is based on the view that each workplace itself represents a production system and enables value creation through the use of production resources. The workplaces are bound to a system-specific organisational form and are themselves a part of an overall production system. That is, as long as it does not represent the factory itself or a part of another, hierarchically higher-ranking system such as the line or segment. Based on this type of analysis it can be concluded that each production system can contain multiple subordinate production systems. The implementation of this fact is done with the aid of a modelling approach of object orientation (see Sect. 6.1.3), where the main class of

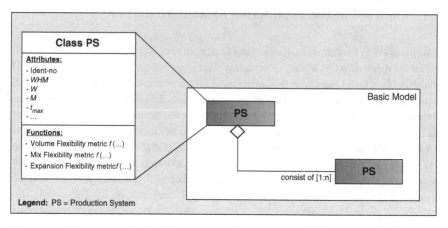

Fig. 3.12 Reference model of the production system model with the class "production system"

the reference model is the class called "production system". It describes an identifier in the form of attributes used to uniquely identify a production system as well as the parameters used as basic input data for the flexibility calculation, as shown in Table 3.2 (p. 57), Table 3.3 (p. 57) and Table 3.4 (p. 58). In addition to the attributes, this class also provides functions. They correspond to the metrics used to calculate the volume, mix and expansion flexibility in production systems.

The following Fig. 3.12 once again illustrates this relationship, drawing on the object-oriented modelling language UML (see Sect. 6.1.3.1).

3.4.2   The Principle of Inheritance of the Reference Model

The principle of inheritance is based on the object-oriented paradigm which allows, apart from the large freedom of movement in the parameterisation of the reference model, the formation of new classes that can be hierarchically constructed using the main class "production system". Similar to the observation level presented in Sect. 2.1.3, four sub-classes "factory", "segment", "line" and "workplace" will also be created. They include the inherited attributes and functions of the main class, but can however be complemented by additional attributes and functions depending on the information demands for the flexibility determination. As shown in Fig. 3.13, an additional feature of the class "workplace" for example, can be the calculation of production capacity in order to obtain information on how much spare capacity remains at a workplace at break-even production.

As a result of the inheritance options provided by the reference model, a user can also deviate from the hierarchy shown in Fig. 3.13 by generating new classes. They can be easily connected with the flexibility metrics, given the stipulated attributes

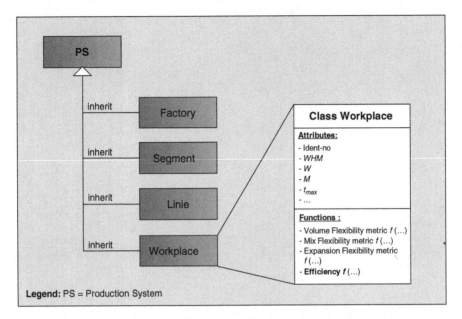

Fig. 3.13 Inheritance principle of the reference model

and functions. Thus, the development of a PSM for example would also be feasible, which due to the specific structure of the production system to be evaluated, unites various workplaces and, where necessary, lines which themselves belong to a segment.

This example shows that the PSM represents a specific characteristic of the object-oriented reference model, which is universally adaptable to the respective production system.

Chapter 4
Practical Experience in the New Methodology

With the goal of verification of the evaluation methodology, a software tool called ecoFLEX was developed, which makes use of the mechanisms for the assessment of flexibility of production systems described in the previous chapter. The detailed description of this software tool in terms of its architecture, its implementation and its operating principle will be the first topic discussed in this chapter. Since a software-based realization of the methodology alone is not sufficient to prove its viability in real conditions, the software tool was practically tested subsequent to its development through an industrial case study. The second focal point in this chapter is therefore the detailed explanation of the resulting experience from the analysis of Quantity, Mix and Expansion flexibility. From this the evaluation methodology can be verified a by considering a detailed review of all requirements and the fulfilment of these.

4.1 Implementation of the Evaluation Methodology

A fundamental task in the realization of the ecoFLEX tool was the implementation of a suitable software (in accordance with Sect. 2.5.4), the details of this which follow here. The software was developed based on modern technologies to achieve various technical advantages, such as simple advancement opportunities or higher versatility e.g. in distributed environments. This in turn results in ergonomic benefits for both the user and for the developers of the ecoFLEX software.

4.1.1 Software Architecture of ecoFLEX

The development of ecoFLEX focused on providing the necessary functionality required for the automation of the procedure for the flexible evaluation of production systems, as described in Sect. 3.2. With this in mind, a software concept was developed, intended to support three successive base iterations (see Fig. 4.1).

S. Rogalski, *Flexibility Measurement in Production Systems*,
DOI 10.1007/978-3-642-18117-7_4, © Springer-Verlag Berlin Heidelberg 2011

Base Iteration 1:

Construction of a production system model as an abstract representation of the production infrastructure which is under investigation

Base Iteration 2:

Collection and structuring of relevant data for flexibility calculation, based on the chosen evaluation objects/abstraction

Base Iteration 3:

Flexibility calculation through the application of the flexibility metrics to the evaluation objects of the production system model

Fig. 4.1 Base iterations for flexibility evaluation, automated through ecoFLEX

An aspect which until now was not supported by software was the automatic detection of flexibility deficits and a concurrent generation of the proposed solution for their closure. This *Base Iteration 4* (flexibility interpretation/evaluation) is, at the present stage of eocFLEX development, reserved only for the user. The Fig. 4.2 shows the concept of the modular software architecture of ecoFLEX, which ensures the highest possible degree of user support in the flexibility assessment.

The basic elements of the architecture form the interfaces; the configuration module PSM; the module for data processing; the module for the flexibility assessment; the graphical user interface (GUI) and the database. These terms are explained below:

- *Interfaces:* Since the vast majority of data needed to conduct the flexibility assessment is already available in the different operational planning and control systems, the appropriate interfaces to these systems are necessary. As seen in Fig. 4.2, there are two basic types of interfaces. One type being the interfaces for collecting flexibility relevant data from the operational systems in a manufacturing environment. These can be used to implement planning and controlling related functions of short-and medium-term production and commercial operations. Prime examples are ERP systems, which cover products which possess such functionality. PPC- BDE-, SCM- systems or other systems for controlling and scheduling also partially contribute to the support of the overall functionality and therefore provide necessary flexibility information. The other type is made up of the ecoFLEX interfaces to digital tools usually used for long-term factory planning. These are used to transfer object relevant information on the production infrastructure to ecoFLEX in order to build the PSM. This information is obtained from the digital models for the layout plan and the line- and workplace configurations.

- *Configuration module PSM:* This module directly supports the first of the three base iterations listed in Fig. 4.1, namely the construction of the production system model. Using a configurable set of standards especially provided for this function, the hierarchical structure of the production system can be automatically generated. The structure is derived from the extracted object information from the CAD models built in the digital factory planning system. Modeled

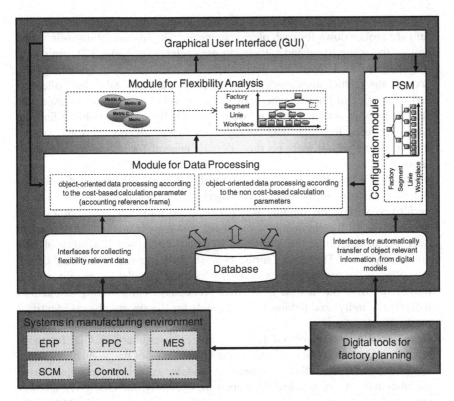

Fig. 4.2 System concept of software architecture of ecoFLEX and their integration into the IT landscape of the user companies

system objects like workplaces, lines and segments as well as their mutual dependencies can be identified due to the semantics contained within the standards. The semantics refer to the CAD Plant Design symbols from each block name, label, layer, line type etc. As a result, the corresponding information based objects can be compiled. Provided that an appropriate demand for manual rework exists and that automated model building is not possible, the PSM configuration module provides the relevant features which are also suited to the building of scenario alternatives.

• *Module for Data Processing:* This module forms the basis for the second of the three base iterations, the collection and structuring of relevant data for the flexibility analysis. A prerequisite for this is the PSM derived from the configuration module, since this can clearly define the system objects and their mutual dependencies. This results in a purposeful, object related data mapping. We distinguish between two types of data structure, which is why each has a separate sub-module available. The first is the object-oriented data processing according to the cost-related calculation parameter as shown in Table 3.3. The extracted cost information under the cost accounting reference frame presented in Sect. 3.3 is hereby separated into variable and fixed cost components, which in turn are broken down for each object parameter. The second sub-module which defines the non-cost-related calculation parameters (see Table 3.2), deals with the assignment of the object data which fall into this category, taking into account the relationships modeled in the PSM. Fundamentally, the collection and structuring of both categories of flexibility relevant data on the respective sub-modules is automatically, as well as manually possible. This offers the advantage of a thorough data survey even in the absence of some system interfaces and allows user-specific value assignment for alternative scenarios, as well.

- *Module for Flexibility Analysis:* The third base iteration, the application of flexibility metrics to the evaluation objects of the production model, will be performed within this module which is divided into two sub-modules. The first deals with the implementation of the assessment methods described in Sect. 3.2 to determine the Volume-, Mix- and Expansion flexibility of production systems. The actual application of these methods, however, comes into play in the second sub-module in which the calculation parameters are transferred from the module to the data processing. The resulting calculation delivers the individual flexibility coefficients for the system objects identified by the PSM.

- *Graphical User Interface (GUI):* The graphical user interface, GUI for short, has two fundamental tasks. One being the visualization of the production system model created in the PSM configuration module, as well as the object based calculation results of the flexibility analysis. Secondly, the GUI gives the user the possibility to exercise a simple influence on the structure of the PSM to be evaluated and its accompanying data allocation, because it allows the interactive access to the PSM configuration module and the module for data preparation (see the preceding description of the two modules).

- *Database:* The database provides the foundation for the persistent, model-related filing of all the information on a production system needed for the flexibility analysis. All of the collected data and structures can be permanently stored in the database, enabling, for example, comparisons between different production systems or different developmental statuses of a system. Data from the internal information management of ecoFLEX are also stored here, such as semantic information to CAD Factory Design symbols which are used in the PSM configuration module standards.

4.1.2 Implementation of ecoFLEX

The basis for the implementation of the evaluation methodology was the open development platform *Eclipse*. The result represents the flexibility evaluation tool ecoFLEX, the development of which involved the following software development technologies:

- The implementations were performed in version 3.4.0 of the *Eclipse Development Environment*, using the Java programming language based on the *Java Development Toolkit* (JDK) version 1.6_10.
- The *Rich Client Platform* (RCP) of Eclipse (Eclipse RCP) version 3.4.0 was used for the construction of the Graphical User Interface.
- The relational database *MySQL* version 5.1 was utilised for persistent data storage and data management. This allows the storage of ecoFLEX objects and their associations using the Java-compatible Open-Source-Persistence-Framework *Hibernate* version 3.3.1.

Figure 4.3 shows a screenshot of the window of the Eclipse development environment. The Project Navigator can be seen on its left side (1), which contains the project folder ecoFLEX. This includes, inspired by the modular structure of the developed software architecture (see Fig. 4.2), various subfolders in which the project files generated for the prototype compilation are stored. They in turn describe the specific ecoFLEX functionality in the form of source code stored there, as shown in the right side (2) of the Eclipse window. Shown there is the Eclipse

Fig. 4.3 Project structure of ecoFLEX in the Eclipse development environment

workspace with an excerpt from the source code of the file *MengenFlex.java* from the subfolder *Evaluation method*. The structure of the source code in the project files can be extracted from the *Outlines* in the Project Navigator in the Eclipse window (3).

As mentioned above, the Java programming language forms the basis of the implementations which were undertaken. A decisive reason for this is, for one, its object orientation which accommodates various mechanisms in the evaluation methodology (see Sect. 3.2). For another, the architecture neutrality is preserved, which is beneficial for the ease of integration of ecoFLEX into its intended IT infrastructure. Multiple project types can be distinguished for the Eclipse Development Environment which was implemented in this context and in the interest of a progressive, supportive implementation approach.

They are characterized by different development goals (such as Database applications, Internet applications, Eclipse Plug-ins or Rich Client applications) and therefore by different source file types, as well as their applicable tools and Plug-ins. For the development of ecoFLEX, the project type *Plug-in Project* (RCP) was selected, for which the RCP- Framework of Eclipse was used to ensure a modular, simple and clear configuration of the graphical interface. Another framework that ecoFLEX uses is *Hibernate*. It creates an object-relational bridge to the database, and can thereby store the status of an ecoFLEX object in the MySQL database and can also recreate it.

4.1.3   Functionality of ecoFLEX

In order for the ecoFLEX software to be employed in its basic configuration, a suitable computer system is required. The minimal requirements of such a system are the existence of 32-bit computer architecture with a processor capacity of 1 GB and a working memory capacity of 512 MB, as well as having Windows, Linux or Solaris as operating systems to support the required Java console. After installation of the ecoFLEX software, it may be started via a specially provided program icon in the program folder. The subsequent approach to the targeted flexibility analysis is described in the following sub-chapter.

4.1.3.1   Structure of the Program Window

As ecoFLEX starts, a special programme window will open which is divided from a functionality perspective, into an analysis area and parameter area (horizontal view). Regarding the flexibility analysis, these two areas are divided into an original- and an alternative representation (vertical view), as can been seen in Fig. 4.4.

The *analysis area* itself consists of three so-called functional fields. On the left in the original view associated field (1), the model of the considered production system is statically presented in its initial/original configuration with the associated

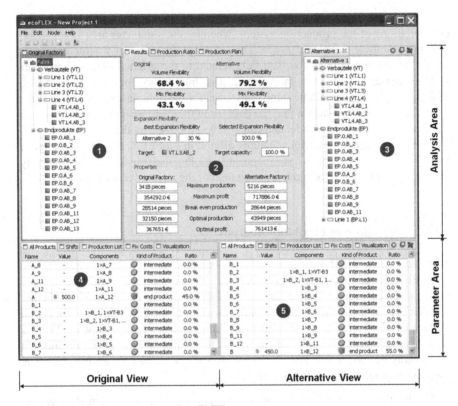

Fig. 4.4 Graphical user interface of ecoFLEX

system objects. On the other hand, configurations which differ from the original model can be compiled in the alternative view associated field (3). Here, in contrast to the field (1), a unique toolbar with the appropriate functionality for model changes is integrated. The middle field (2), which includes both an original and an alternative view, consists of three tabs. The first, called *Results*, allows the display of object-oriented indices to the original as well as to the alternative models. This does not only concern indices for Volume-, Mix- and Expansion flexibility of the different system objects, but also production management parameters such as break-even points or maximum capacities. The latter are regarded as so-called auxiliary parameters, which help to better assess the individual object-flexibilities. The second tab, *Production Ratio,* displays the product mix which underlies the flexibility calculations and the product mix for optimal production. The third tab, *Production Plan,* lists the various system object related production plans with their specific workloads.

All the required data for the calculation of the indices are encompassed in the *parameter area* of the ecoFLEX programme window. It is also possible to create a visualization of the CAD (simulation) models, built in the digital factory planning, of the system objects collected in PSM. This generally leads to, as in the analysis area, a dissociation between the original configuration and alternative configurations of the production system, whereby the parameter area is separated into two identical *functional fields*. They contain, in terms of a structured data collection, corresponding tabs. The field related to the original view (4) provides all the necessary computation- and visualization information for the original model, which cannot be changed. The same is also true for the right field (5), the alternative view, but with the distinction that the data listed there are assigned to the different variations of the original model and can also be processed by these variations.

In order to access the various features offered in ecoFLEX, the upper chapter of the programme window also has a menu and a toolbar and provides further logically arranged, smaller toolbars in the function field of the alternative view (see Fig. 4.4).

4.1.3.2   Procedure of Flexibility Analysis

With ecoFLEX, flexibility investigations executed on production systems are generally referred to as *Analysis Projects* and are also stored as such. Intermediate backups of each processed state can be done independently of user activity, whether the user is concerned particularly with the creation of the PSM, the data allocation or the analysis. To create an Analysis Project the three basic steps shown in Fig. 4.1 are followed, starting with the structuring of the PSM (*1st base iteration*) through the "Configuration module PSM" of the software tool ecoFLEX (see Fig. 4.2). The user has two options available to him:

- Firstly, the user may access the digital factory planning system through the interfaces provided by ecoFLEX to automatically transfer the stored CAD models of the production system into the PSM-specific structure. In addition, the user has to call the *Import* function in the *File* menu, through which the ecoFLEX supported files can be accessed using the Wizard. The generated PSM is then represented in the alternative view in the right field (3) of the analysis area (see Fig. 4.4) and can then be manually re-processed.
- The second option is to build the PSM entirely by hand, which is especially necessary when digital factory models are incomplete or not available at all. This is performed in the alternative view of the analysis area, where the necessary functions for model construction can be found.

Once the PSM has been fully created in ecoFLEX, the system objects required for the flexibility calculations are assigned the necessary data (*2nd base iteration*). Once again the user has the choice of an automated or manual procedure:

- For the automatic execution of the second base iteration, the *Import* function within the *File* menu is used. With the support of the Wizard, the user can access the systems which are coupled to ecoFLEX from the operational production environment. These are then sorted for flexibility relevant data by the "Data Processing Module" (see Fig. 4.2), formatted and allocated to the system objects contained within the PSM. All the data interpreted in this way can later be displayed in the parameter area of the alternative view (see Fig. 4.4) as normalized within a uniform evaluation period.[1]

- In case automatic data imports are not feasible or if re-processing is necessary due to incorrect data interpretation, ecoFLEX also supports manual data capturing. The relevant Wizard-based functions are available in the parameter area of the alternative view and can be run directly from the parameter area (see Fig. 4.5). It is important to note that the input data is normalized over a standard evaluation period.

Example. To make the procedure of manual data mapping more understandable for the reader, Fig. 4.5 illustrates an ecoFLEX screenshot that demonstrates the capture of flexibility relevant data using the example production system from Sect. 6.4. It shows that the first step (1) allows for the selection of the system object, thus enabling the already allocated object data to be viewed via the tabs in the parameter area. Workplace *AP1.1.1* can be seen in the screenshot. The workplace is allocated a "manufactured on-site" product, *Product 1*, with its object-specific properties such as production time, scrap rate, variable costs etc. To allow this data to be modified or new data to be collected in the second step (2), the corresponding icons which start the relevant Wizard are made available in the toolbar. The wizard "Add Product" shown in the screenshot illustrates how to add an additional product, *Product 2*, which is assigned to workplace *AP1.1.1* along with its data necessary for the flexibility analysis.

The entire procedure can be carried out successively for all other system objects in the PSM, the processing is however limited exclusively to the alternative view.

Once the PSM and the flexibility relevant data of a production system have been completely represented in ecoFLEX (i.e. the first two base iterations have been completed), the last of the three base iterations, the calculation of the object-related flexibilities, is prepared (*3rd base iteration*). In addition, the data represented in the alternative view are transferred to the original view and at the same time the required Product mix necessary for the flexibility analysis is established. The function "Set Alternative as Original" in the "Edit" menu is provided for this purpose, through which the calculation of the various object-related flexibility parameters is automatically executed via the "Module for Flexibility Analysis" (see Fig. 4.2). The same happens during the scenario observations, in which flexibility calculations are immediately updated with any revised value-assignment.

In conjunction with an update of the flexibility indicators, the additional auxiliary parameters are also recalculated (see field (2) in Fig. 4.4) which supports the user to quickly identify any flexibility deficits and to search for suitable alternative solutions. The specific procedure is discussed later in Sect. 4.2.3.

---

[1]The evaluation period represents a sample average period in which all the parameter data necessary for the flexibility calculations (cost-related and non-cost-related) are normalized to a predefined time interval e.g. weekly or monthly intervals (see also Sect. 3.3.4.2).

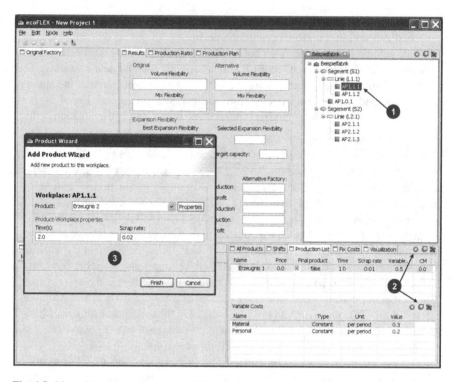

Fig. 4.5 Manual data capturing in ecoFLEX

4.2 Case Study-Related Application Experience

The models depicted thus far for the development and implementation of the evaluation methodology can be traced back to problems experienced in practice (see Sect. 1.2). It must also be investigated whether the methodology is suitable for its intended use. For this reason, a real production system of a series producer of infotainment systems[2] will be used as a case study to assess the methodology with respect to the flexibility metrics and the production model. Due to an existing confidentiality agreement between the author and the user company, the reader will only be privy to normalised information.

---

[2]The word "infotainment is formed from the two words information and entertainment and describes the link between the imparting of information and entertainment. An infotainment system makes use of a so-called Multi Media Interface and can be used for example in motor vehicles. In addition to the radio programme other complex tasks are also available, such as an instant, written reports of the current traffic situation, as well as the displaying of the map navigation system while simultaneously operating the phone and/or comfort features of the vehicle.

4.2.1 Initial Situation at the Company

The production company assigned with the verification is internationally positioned and employs more than 7,000 employees at approx. 30 locations around the
world. Its core business includes the development, production and sales of infotainment products, primarily for well-known automobile manufacturers. Besides
the OEM-business segment, the company also develops and manufactures their
own products. Due to growing competitive pressure in the world market, the
demand for high competitiveness is also increasing. Large variety, low cost in
conjunction with a high degree of innovation and quality as well as shortened lead
times with equally short notice periods for orders, lead to great uncertainty in
planning. This uncertainty is worsened by the global financial crisis which, since
mid-2008 has greatly affected the automotive sector. Thus, the company is forced
to accurately estimate its commercial trade to be able to respond adequately
and flexibly to any unexpected orders. In this context, the question arises as to
possibilities for the assessment of degrees of freedom of the company-specific
production system, and to what extent:

- This can react to changing demand levels (Volume flexibility)
- The changes in the composition of Product-/Variant mixes affect its profitability
  (Mix flexibility)
- Its capacity can be extended to deal with production bottlenecks and which
  alternative courses of action can be resorted to in order to do so (Expansion
  flexibility)

As a result, the company has already begun to use digital factory planning tools,
to ensure a higher degree of predictability. Due to the resulting virtualization
potential, it was possible to simulate various processes in production planning and
production, so that important cost and time savings could be achieved while still
increasing the quality of the planning. A weakness was however observed in the lack
of an appropriate evaluation basis for the quantification of flexibility margins within
the production infrastructure. Hence, investigations to determine the correct degree
of flexibility are only possible by so-called "rehearsing", which is why flexibility
deficits are usually separately identified and evaluated. A targeted approach to the
comprehensive governing of these deficits in economic terms is hardly ever accomplished through this, since there are no suitable quantifiable parameters.

4.2.2 Object Area of the Case Study

The objective of the case study was, based on the company's situation as outlined
above, to test the evaluation methodology through the implementation of ecoFLEX
at a German production location. The test environment was the production system
of a factory, in which two variants of a certain infotainment product for vehicle

installation for a major automobile manufacturer are produced. The applicable sales prices are regulated by the appropriate framework agreements, to ensure clarity on the prices with the attainable quantity unit revenue. However, there is a large uncertainty regarding the order quantities and the variant mix. The framework agreements do however guarantee a minimum purchase order, which is why certain assumptions can be made with regards to the Variant mix. However, the actual production quantities and their retrieval dates cannot be predicted. Which poses a major challenge for the production management in terms of economic manufacturing because the production system must retain a sufficient level of flexibility.

The production of two variants of the said infotainment product, hereafter referred to as variant A and variant B, in the observed factory area is divided into two segments, one of which is intended for the preliminary production and the other for the final production. Both segments exhibit both manual and semi-automatic workplaces that follow the so-called in-line configuration production principle. There is therefore no direct time synchronization between them, so that the corresponding buffers are available in the form of container boxes. The thus induced discontinuous material flow is controlled by an automatic transport system. The procedural set up of the two segments is as follows (see Fig. 4.6):

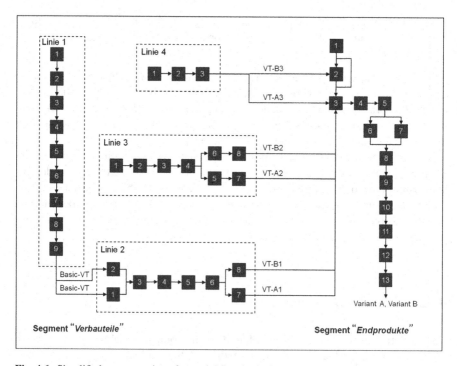

Fig. 4.6 Simplified representation of material flow in the investigated area of the user company's factory

- The first segment called Verbauteile deals with the production of the necessary preliminary products, hereafter referred to as Verbauteile *VT*. These are considered as non sellable. This process makes use of four lines with a different number of workplaces. The *VT.L1* line produces a basic variant called *Basic-VT*, which is required for both option A and option B. This base variant becomes the input for line *VT.L2* which produces both *VT-A1* and *VT-B1*, which in turn are passed on to the segment 2. These are produced partially at single-product workplaces and partially at double-product workplaces. The same is true for the line *VT.L3*, from which both *VT-A2* and *VT- B2* emerge, but without the incorporation of the basic variant. In contrast, the line *VT.L4* deals exclusively with the production of parts *VT-A3* and *VT-B4* at double-product workplaces.
- The workplaces contained in the second segment, called *Endprodukte* do not have any line division, but can be viewed as one large line due to their row-configuration. Here, the Verbauteile created in the first segment are processed together with externally purchased components, which ultimately leads to the production of the two variants which are to be sold. In addition, the Verbauteil *VT-B3* is incorporated into the second workplace of this segment, where it is manually joined with a purchased component which itself is pre-assembled at a preceding workplace. In contrast, the processing of *VT-A1/B1, VT-A2/B2* and *VT-A3* in the "Endprodukte" segment only begins at the third manual assembly workplace.

The Fig. 4.6 shows once again the procedure for material flow of both segments, whose workplaces are marked as numbered squares.

The procurement and installation of one of these workplaces incurs, depending on the type, costs of €1,000 up to €253,000, with the average acquisition costs amounting to around €7,200 per workplace. The amortisation period is 6 years.

Given the existing uncertainty in planning and the resulting demand for flexibility, a variable working hour configuration was introduced. This was based on six different working hour models, which themselves were based on either single or double shift production. A possible third day-shift was not considered, as this time is reserved for maintenance of the workplaces and transportation systems. Table 4.1 shows the different working hour models with their associated maximum operating times (in seconds) per working week and the corresponding daily working hours.

From the perspective of the necessary human resources for the product creation, the normal shift provides a standard team of 18 employees in the "Verbauteile" segment and ten employees in the "Endprodukte" segment. Here, with the exception of four electrically skilled employees required in the "Endprodukte" segment, only semi-skilled workers are employed. In addition, one foreman per shift monitors and coordinates the production processes taking place in each of the two segments. To implement the double-shift, the option of resorting to temporary workers who can be easily integrated into the working sphere of the semi-skilled workers needs to be considered. The electrical technicians who belong to the core workforce are distributed equally between the early and late shifts, while the daily working hours of the foreman are increased by 2 h (from 9:00 to 19:00).

Table 4.1 Working hour models of the considered plant

Working hour model	Description	Daily working hours	Operating time t_{max} per working week
WHM_1	*Normal shift*	8:00–16:00 (Mo–Fr)	144,000 s
WHM_2	*Normal shift – long* (+ 2 h overtime)	8:00–18:00 (Mo–Fr)	180,000 s
WHM_3	*Normal shift – Saturday* (+ 2 h overtime + Saturday)	8:00–18:00 (Mo–Fr) 8:00–16:00 (Sa)	208,800 s
WHM_4	*Double shift*	Early: 6:00–14:00 (Mo–Fr) Late: 14:00–22:00 (Mo–Fr)	288,000 s
WHM_5	*Double shift Saturday* (+ Saturday)	Early: 6:00–14:00 (Mo–Fr) Late: 14:00–22:00 (Mo–Sa)	345,600 s
WHM_6	*Double shift – weekend* (+ Saturday + Sunday)	Early: 6:00–14:00 (Mo–Su) Late: 14:00–22:00 (Mo–Sa)	403,200 s

4.2.3 Application of the Evaluation Methodology in the Context of the Case Study

The production system described above underwent a detailed flexibility analysis, which resulted in a complete representation of all the evaluation-relevant system objects and their associated flexibility information in ecoFLEX. The data sources used were systems of the digital factory planning utilised at the company location, the systems of the Enterprise Resource Planning and the captured production data. They provided current and useful data for the flexibility measurement, their normalization was performed on a weekly basis. The smallest possible evaluation period is made up could be about this all applicable to the commercial sector working hour models, which suited the most precise data collection. A necessary condition for the carrying out of the flexibility investigations was the determination of the quantity ratio for the production of both product variants. This so-called sales-related product mix ratio was determined for the considered production system as a result of an analysis of previously acquired production figures for the current fiscal year combined with the projected revenue expectations for next year. The ratio which emerged for the said infotainment product was 45% for product variant A and 55% for product B. The following Product mix vector was derived from this:

$$v = (0.45 \times MI_A, 0.55 \times MI_B)^T$$

Table 4.2 shows the resulting parameters determined by ecoFLEX for the Volume and Mix flexibility of the various system objects of the analyzed plant. The table shows selected individual flexibilities of the workplace level, as well as line-related indicators, segments and the factory sector as a whole.

Table 4.2 Calculated volume- and mix flexibilities for selected system objects of the production system in the factory

Factory: $F_{Volume} = 68.4\%$; $F_{Mix} = 43.1\%$

System object	F_{Volume} (in %)	F_{Mix} (in %)	System object	F_{Volume} (in %)	F_{Mix} (in %)
Segment "VT"	**68.4**	37.4	**Segment "EP"**	**69.0**	54.2
Linie VT.L1	79.3	0	*EP.0.AB_1*	82.3	54.2
Linie VT.L2	**68.4**	54.2	*EP.0.B_2*	94.9	0
VT.L2.A_1	93.3	54.2	*EP.0.AB_3*	**69.0**	54.2
VT.L2.B_2	90.3	54.2	*EP.0.AB_4*	86.1	54.2
VT.L2.AB_3	83.2	54.2	*EP.0.AB_5*	94.5	54.2
VT.L2.AB_4	89.6	0	*EP.0.A_6*	87.2	0
VT.L2.AB_5	91.1	0	*EP.0.B_7*	81.2	0
VT.L2.AB_6	92.5	54.2	*EP.0.AB_8*	94.6	54.2
VT.L2.A_7	**74.2**	54.2	*EP.0.AB_9*	94.6	54.2
VT.L2.B_8	**68.4**	54.2	*EP.0.AB_10*	91.4	54.2
Linie VT.L3	86.1	54.2	*EP.0.AB_11*	87.5	54.2
Linie VT.L4	86.7	54.2	*EP.0.AB_12*	93.1	54.2
			EP.0.AB_13	93.6	54.2

4.2.3.1 Analysis of Volume Flexibility

According to Table 4.2 above, the entire factory being examined exhibits a Volume flexibility of 68.4%. Within this percentage range it is possible for the production of the two product variants A and B in the given product-mix ratio to react to market demand fluctuations between the break-even volume and the capacity limit, without risking the efficient production of the variants, and without having to alter the element volume and structure of the production. Expressed in actual production figures, this means a production of 289–917 units of infotainment variant A and 354–1,121 units of variant B. It can be interpreted from Fig. 4.7 that the break-even quantity (field 1) is already achieved with the "Normal shift" operation, while the maximum output rate demands the use of the "Double shift- weekend" model (field 2). On closer examination it is apparent that the workplace *EP.0.AP_13*, where both variants A and B first reach their sale capability, has a relatively low utilization rate of 18% and 20%[3] (8% for A, 10% for B). This leads to a very high Volume-specific individual flexibility of the workplace, with 93.6% (see Table 4.2) which is well above that of the overall system, and hence points to flexible deficiencies. This could be the result of a flexibility bottleneck in the factory, or due to the potential flexibility of workplace *EP.0.AP_13* being unnecessarily large. However, since the other workplaces in the production also exhibit significantly higher flexibilities than 68.4% (see Table 4.2), it is assumed to be due to a flexibility bottleneck.

---

[3]In normal shift operation 8% for variant A and 10% for variant B (= 18% total utilisation), in double-shift weekend operation 9% for variant A and 11% for variant B (= 20% total utilisation).

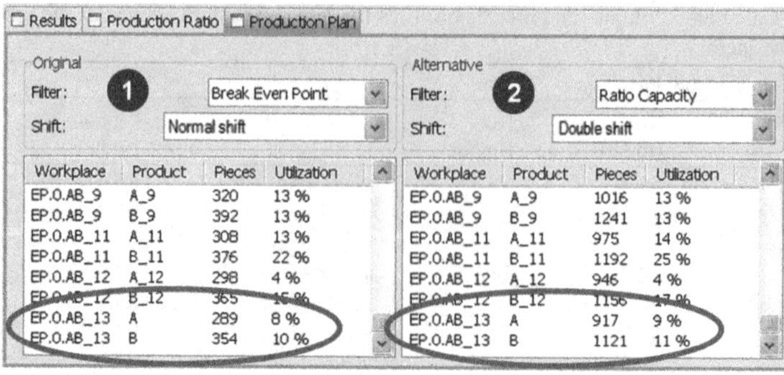

Fig. 4.7 Extract from the analysis of ecoFLEX of the workplace-related quantity and utilization

To determine the flexibility deficit in the factory being considered, the individual flexibilities of the various system objects are to be considered, and should be done so systematically because of the hierarchical dependencies in the production system. This is illustrated in Table 4.2 (p. 123), whereby workplaces with a lower Volume flexibility adversely affect the flexibility of their parent system objects. Thus in terms of the flexibility analysis of the user company, it quickly became apparent that their production system exhibits three serious flexibility bottlenecks. These are both workplaces and *VT.L2.A_7* and *VT.L2.B_8* in the line *L2* of the segment "Verbauteile" with 74.2% and 68.4% affected, while the third is the workplace *EP.0.AB_3* in the "Endprodukte" segment with 69%.

4.2.3.2   Analysis of Expansion Flexibility

Subsequent to the Volume flexibility analysis, an assessment of the identified flexibility deficits showed that they were not due to a bad cost-effectiveness of the product creation at the three workplaces, but could be traced back to a low maximum capacity. The required additional capacity of the individual workplaces could be quantified with ecoFLEX as follows:

- 26% for workplace *VT.L2.A_7*
- 54% for workplace *VT.L2.B_8*
- 51% for workplace *EP.0.AB_3*

According to these findings, it was necessary to search for alternative solutions which could guarantee the necessary capacity demand within a 6 month time period (the sum of deciding- and implementation periods, see Fig. 2.7), in order to eliminate the flexibility deficits in the production system. For the work-place *EP.0.AB_3* the only option was to build a redundant workplace. Further considerations regarding the insertion of additional mechanical supports were however rejected, since manual execution was not effective due to the distinctive nature of the operations performed there. In contrast, there are three alternatives listed below for the two workplaces *VT.L2.A_7* and *VT.L2.B_8*, which only concern the line *VT.L2*:

- **Expansion Alternative 1:** Structure of workplace VT.L2.B_8(new) and VT.L2.A_7(new) as duplicates of workplaces VT.L2.B_8 and VT.L2.A_7
- **Expansion Alternative 2:** Structure of a workplace *VT.L2.B_8(new)*, redundant to *VT.L2.B_8* and workplace *VT.L2.A_7(50% new)*, redundant to *VT.L2.A_7*, however with a reduced efficiency (50% of the existing workplace *VT.L2.A_7*)
- **Expansion Alternative 3:** Structure of the multipurpose workplace *VT.L2. AB_9(new)*, at which both the operations of the workplaces *VT.L2.A_7* and *VT.L2.B_8* are executed

In order to compare the three listed expansion options and thus determine the best one, the Expansion flexibility of line *VT.L2* was considered using ecoFLEX. A line-related target capacity of up to 23,641 parts is assumed. It is derived from the largest required additional capacity from within the line. This affects the workplace *VT.L2.B_8* with a required increase in the maximum capacity of 54%. Table 4.3 shows the resulting alternative-dependant calculation results for the Expansion flexibility of the line *VT.L2*, where the underlying costs and non-cost-related calculation parameters are given in the Sect. 6.5. Consideration of an alternative for workplace *EP.0.AB_3* was not necessary, since as already mentioned, only the construction of a redundant AP was to be considered.

Using the parameters determined for the Expansion flexibility it can be seen that the construction of a multi-purpose workplace (Alternative 3) is desirable. The calculated index of 116.1% implies an improved Volume flexibility in line *VT.L2*. Therefore, it can be as viewed as "fully expansion flexible" (see case distinctions in Sect. 3.1.3). Indeed both of the other expansions alternative display a higher maximum capacity, which could easily persuade one to give them preference. However, that would lead to the construction of unnecessary flexibility potentials, which are regarded as less economical, as they only reach the break-even amount for the line flexibility efficiency later (see Table 4.3).

Table 4.3 Indicators for expansion capacity of line "VT.L2" (alternative-dependant)

Systemobject: Line "VT.L2"	Break-even-quantity (parts)	Maximum-capacity (parts)	Target-capacity (parts)	Expansion flexibility
Expansion alternative 1 Structure of workplace *VT.L2.A_7 (new)* and *VT.L2.B_8(new)*	5,602	28,821		111.5%
Expansion alternative 2 Structure of workplace *VT.L2.B_8 (new)* and *VT.L2.A_7(new 50%)*	5,109	28,144	**23,641** (+54%)	114.6%
Expansion alternative 3 Structure of the multi-purpose workplace *VT.L2.AB_9(new)*	4,867	24,409		116.1%
Original configuration	4,849	*Maximum capacity:* 15,351 parts		

The expansion of production systems at the two workplaces*VT.L2.AB_9(new)* and *EP.0.AB_3(new)* leads to an increase in the Volume flexibility to 79.1% (up 10.7%) and the Mix flexibility to 49.1 (plus 6%). This indicates a clear improvement in comparison with the existing system configuration.

4.2.3.3 Analysis of the Mix Flexibility

As presented in Table 4.2 (p. 123) from the introduction to the case study analysis, the studied production system has a total of 43.1% Mix flexibility, which represents a considerable risk of an economical production with product mix fluctuations. It was demonstrated using ecoFLEX that a loss in demand for product B would result in an 80% fall in profits. A Stop in the demand for the product *A* instead of *B* would have even more dramatic repercussions, because the profit potential would fall by about 95%, which also affects the Volume flexibility of the system. Figure 4.8 shows the analysis area of the ecoFLEX window from which the changes in the Volume flexibility as well as the break-even and the maximum production volume emerge, as a result of the stop in production of variant *A*.

The reason for the bad Mix flexibility of the considered factory is the high number of single product workplaces, which include 20 of the 41 workplaces which are distributed in equally over both product variants. Due to the associated purchasing costs, they assume relatively high fixed costs in the production system, which is

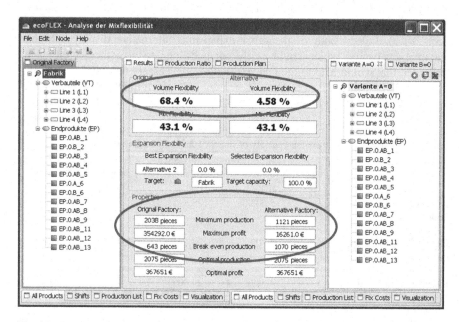

Fig. 4.8 Extract from the analysis area of ecoFLEX to study the effects of non-production of the product variant A

only later covered by the sales income. In the case of the non-production of either of the two variants, these fixed costs remain while the income falls away. Covering the fixed costs is therefore only possible from sales of the other product variants, as are the costs for the unutilised workplaces.

A fall in demand for variant *A* poses a higher economic risk. Therefore, it can be accepted within the production system to be the more valuable product. However, this should not be reason to neglect variant *B*, because due to the especially purchased individual workplaces, a maximum production success is only attainable with the production of both variants. The analysis performed by ecoFLEX therefore gave the following product mix ratio to reach a maximum profit in terms of optimal production:

- 51% of the total amount of variant *A*
- 49% of the total quantity of variant *B*

In Table 4.4, the output rates for different product mix ratios for variant *A* and variant *B* are compared.

This table shows that the factory is relatively well set out with its current configuration designed for the product mix ratio of 55% to 45%, since the loss in profits compared to optimal mix is kept within limits. Nevertheless, this fact does not change the bad Mix flexibility, applicable to the entire system. In order to find possibilities for improvement, the already identified Mix flexibility indicators for the various system objects were more closely examined by ecoFLEX (see Table 4.2). This showed that the segment "Verbauteile" with its line *VT.L1* (Mix flexibility = 0) had an especially negative impact on the overall impact of Mix flexibility of the system because it only makes use of single product workplaces.

A reasonable solution option that was identified in the case study to address this Volume flexibility deficit, involved the closure of the line *VT.L1*. Although this means the intermediate product built there, *Basic-VT*, would have to be outsourced, the Mix flexibility would rise from 43.1% to 54.2%. Thus, the economic risk of compromising the success of production due to changes in the product-/variant mix could be lowered by 11.1% points. Moreover, there would be no fear of a negative impact on the Volume flexibility, as long as the cost of procurement of *Basic-VT* did not exceed 11 Euro.

Table 4.4 Comparison between production characteristics of different production scenarios

Parameter	Assumed product mix	Optimalmix	Non-production A	Non-production B
Product mix ratio	Variant A = 45%	Variant A = 51%	Variant A = 0%	Variant A = 100%
	Variant B = 55%	Variant B = 49%	Variant B = 100%	Variant B = 0%
Break-even-quantity (in parts/period)	Variant A = 289	Variant A = 324	Variant A = 0	Variant A = 757
	Variant B = 354	Variant B = 314	Variant B = 1.070	Variant B = 0
Max. production quantity (in parts/ period)	Variant A = 917	Variant A = 954	Variant A = 0	Variant A = 1,121
	Variant B = 1,121	Variant B = 1,121	Variant B = 1,121	Variant B = 0
Max. attainable profit (in Euro/periods)	€354,292	€367,651	€16,261	€74,178

Table 4.5 Comparison of the flexibility parameters between the original configuration and the identified improvement measures in the considered production system

Parameter	Original configuration	Construction of workplaces VT.L2.AB_9(new) + EP.0.AB_3(new)	Closure of VT.L1 and construction of workplaces VT.L2.AB_9(new) + EP.0.AB_3(new)
Volume flexibility	68.4%	79.1%	79.8%
Mix flexibility	43.1%	49.1%	57.3%

In conjunction with the findings of Sect. 4.2.3.3, a system expansion of workplaces *VT.L2.AB_9(new)* and *EP.0.AB_3(new)* is desired, together with outsourcing of the intermediate product *Basic-VT*. This as the sole improvement measure would, in comparison to the construction of the two new workplaces, increase Volume as well as the Mix flexibility of the factory, as Table 4.5 shows.

4.2.3.4 Appraisal of Flexibility Analysis

The evaluation method in the form of ecoFLEX was proven to be very successful in the practical trials. Through the method the flexibility of the investigated production system could be quantified and assigned individual system objects, which formed the basis of a detailed flexibility analysis. From the perspective of the production planners and other ecoFLEX users within the user company, the software allows a rapid identification of flexibility vulnerabilities which are then correctly classified and purposefully eliminated. This was demonstrated by the flexibility deficits identified in the workplaces *VT.L2.A_7*, *VT.L2.B_8* and *EP.0.AB_3* (see Sect. 4.2.3.1).

It was already suspected that the latter workplace could present a flexibility bottleneck due to its high loading. On the contrary, the assessment of the other two workplaces, in terms of their impact on assessed the flexibility of the overall system, was wrong. This finding surprised the production management of the company, because in spite of the manufacturing simulations in the digital factory planning system, no information was available on this. With the help of the evaluation procedure for Expansion flexibility the best of the detected improvement alternatives could then be easily determined by comparisons of key figures (see Sect. 4.2.3.2), which was considered as a very efficient method. The results of the Mix flexibility investigation were also positively by the people involved. It resulted in a change in consciousness of the responsible management regarding the future of factory planning to minimize the risk of dependence of the production system on the production of specific products or product variants. In addition, the proposal to close line segment *VT.L1* in "Verbauteile" (see Sect. 4.2.3.3) as an impulse generator, using outsourcing measures to improve the system flexibility was also pondered.

Despite the success in the application, the preceding data transfer from the ERP and BDE-Systems was only partially satisfactory. The evaluation-relevant system objects along with their structural and dependency information were able to be extracted from the digital factory planning system and completely mapped in ecoFLEX using the configuration module PSM (see Fig. 4.2). The import of production-related data to the system caused larger difficulties. The reason for this was, irrespective of the previously calculated interface specifications, the lack of integration of the software tool in operational systems in production. This was to be avoided in terms of the first case study investigation at the request of management at the user companies. Instead, these data were collected partially automatically and partially manually with several Microsoft Excel files and then automatically transferred from ecoFLEX. This did however have the adverse impact that due to the lack of a consistent, redundant and media-breach free data storage in the factory, a high level of maintenance was required. Thus, there was a considerable need to evaluate already collected data in terms of their correctness, in case e.g. they were stored in different systems with different values or reading errors occurred. Therefore slight deviations of the values calculated with ecoFLEX from the actual values cannot be ruled out, because the quality of the evaluation results always depends on the quality of the input data.

In general, the evaluation methodology was well accepted by its users, despite the difficulties in data acquisition. The key success factors are based on the one hand, on the software support of the three base iterations for the systematic approach to flexibility investigations (see Fig. 4.1). Through this the time required for the collection of the required data and for the system flexibility evaluation itself was significantly reduced. On the other hand, the positive response was thanks to the analysis options in ecoFLEX. Examples of this were the quick and easy identification of flexibility related vulnerabilities within the manufacturing system as well as the comparison of alternative solutions. Hence the monetary benefits of an ecoFLEX supported flexibility analysis could be made clear to the responsible decision-makers fort his case study.

4.3   Requirement Related Assessment of the Evaluation Methodology

Even if the practical experience of the evaluation methodology presented in the case study was viewed in the industry as very positive, it still required a detailed review of the fulfillment of the requirements. Only this provides binding evidence of target-conforming development and applicability of the methodology, so that in the future many production companies can benefit from the resulting advantages. The following therefore discusses in detail the meeting of the defined requirements, however, the time line of the evaluation proceeds in reverse order to the require-ments definition in Sect. 2.5. Therefore, the verification starts with the software tool and continues through the production system model and the flexibility metrics, up to basic usability.

4.3.1   Verification of the Software Implementation

The concern of the software implementation of the evaluation methodology was to reach a effortless use of evaluation methodology. This gave rise to the software ecoFLEX, which has been subjected to extensive practical testing prior to its use, to verify its compliance with the requirements defined in Sect. 2.5.4.

With the aim of *easy integration*, ecoFLEX includes corresponding interfaces for the inclusion of the developed methodology into the existing IT infrastructures. For this purpose, an XML and an AutoCAD interface are currently available. Using the XML interface, production data from operational systems, such as PPS, BDE, or ERP are transferred to ecoFLEX, provided the relevant adapters are available. Test imports of several hundred, specially created dummy records were handled in ecoFLEX without any errors. In a similar manner the reading from CAD drawings in the formats *.dxf and *.dwg were supported using the AutoCAD interface. Relevant functionality tests were performed with up to 20,000 dummy objects, created by a specially prepared algorithm and stored in DXF files. The read Auto-CAD drawings were converted into XML documents, where the geometric configuration of the drawing objects only provided information relevant for the construction of the PSM, for example, necessary system objects, line affiliations of workplaces, material flow chains, etc. These test scenarios also proved to be extremely satisfactory, since all the relevant drawing components and connections were able to be generated in a fully automated fashion (in accordance with **Requirement R4.1**). The following Fig. 4.9 shows the result of the export of an AutoCAD object of type "workplace" in an XML element and its associated attributes.

In the interest of the *simple and intuitive operability* of ecoFLEX, a graphical user interface was implemented. The ecoFLEX-specific functionality can be accessed via the logically arranged toolbars and integrated menu bar, to conduct a flexibility rating of an industrial production system (see Fig. 4.5). The design and operational instructions of the user interface result from the experience of several test subjects (including those from the production environment) during the development. Thereby, in association with the three previously described base iterations (see Fig. 4.1), improvements were continually made such as Wizard-supports, Tool tips to provide further information about GUI objects, or Drag and Drop

```
<AcDbWorkplaceReference
WPName="EP.0.AB_3" Layer="ID_FFU_06_12_alt"
MaxX="92163.29774753435" MaxY="14157.15433610239"
MinX="90968.29774753435" MinY="13562.15433610239"
PositionX="90965.79774753435" PositionY="13559.65433610239"
PositionZ="0"
Rotation="0"
ScaleFactorX="1" ScaleFactorY="1" ScaleFactorZ="1"
/>
```

Fig. 4.9 Example result of the export of an AutoCAD object

mechanisms. Recent research has revealed that subjects required an average training time of 1 h until a user with a production background was incorporated into the software, and able to independently analyse the example production system presented in Sect. 6.4 using ecoFLEX (in accordance with **Requirement R4.2**).

The key success factor for the *simple and clear representation* of objects in ecoFLEX stems from the breakdown of the graphical user interface into a parameter and an analysis area and their subsequent division into an original and alternative view. This allows the visualisation of a production system within the programme window together with its alternative configurations, which promotes the identification and elimination of flexibility deficits. It draws on both the different system-object-specific parameters for the flexibility analysis, as well as their associated input data used to calculate them. In this context, extensive interviews and tests were carried out with selected users in the ecoFLEX development phase. This resulted in a host of visualization and evaluation requirements, which ultimately led to the user interface with a division for function fields and tabs, illustrated in Fig. 4.4. Upon renewed questioning of the pilot users, a high degree of support was found (in accordance with **Requirement R4.3**).

Through the use of the persistence framework "Hibernate" in connection with the relational database "MySQL", the stipulations for a *persistent storage of flexibility investigations* could be met. As demonstrated by several test storages on which, amongst others, the production case study was based. This resulted in a separate storage of the system in the analysis project "Performance Test" in the database. The performed expansions in the form of degenerate alternative configurations with up to 20,000 system objects, including the associated object information, could not detect any significant performance losses in the ecoFLEX software tool. All object representations, input and calculation updates could attain the retention rate in all test scenarios of under 1 s, error free and without serious delays, despite the extensive volume of data, (in accordance with **Requirement R4.4**).

Although until now there has been no need for functional enhancements of the implementation of ecoFLEX, like for example the addition of further flexibility metrics or interfaces to communicate with other systems, it is still assumed that the emerging additional implementations do not impact on the all of the existing ecoFLEX implementation, but only affect individual modules. This is supported by the modular structure of ecoFLEX (see Fig. 4.2) as well as the structured implementation approach within the Eclipse development environment (see Fig. 4.3). This is to ensure, in contrast to monolithic implementations, an *easy expandability* and markedly reduced expenditure for system maintenance (in accordance with **Requirement R4.5**).

4.3.2 *Verification of the Production System Model*

The PSM developed as part of the conceptual phase was found to be very applicable in practice. All necessary object and structural characteristics of the investigated

factory area from the case study were able to be collected in accordance with the
requirements defined in Sect. 2.5.3, and included in the flexibility analysis.

The basis for this was the specially designed reference model (see Fig. 3.13),
through which the various system objects in the factory could be described via
a *uniform and neutral model notation* in abstraction of the analysed production
system. Thus the complete representation of the factory in ecoFLEX, as well as its
alternative configurations with the system objects arranged at different levels was
easily performed, as shown in Fig. 4.8 (p.126) (in accordance with **Requirement
R3.1**).

As a result of the reference model underlying paradigms of object orientation,
emerges the advantage of a *full recording of the valuation relevant system objects
and their allocation relations*. The diverse, system-inherent features such as material flows or hierarchical dependencies can hereby be replicated in a simple manner
and in the required detail. In Fig. 4.10 below, the structure of the analysed
production system from the case study with its relevant system objects is clearly
apparent (in accordance with **Requirement R3.2**).

In the interest of a correct link between the model symbols and the associated
real-world system objects, in order to avoid erroneous or even conflicting data
mappings, *unique object identifiers* were assigned. The system defined for this
provided the following description logic within the context of the case study (in
accordance with **Requirement R3.3**):

- The denoting of the examined factory area by the name *Factory*.
- *Segments* obtained in accordance with the categorization of their product types,
  and as identifiers the abbreviation "VT" for Verbauteile or "EP" for final
  products.

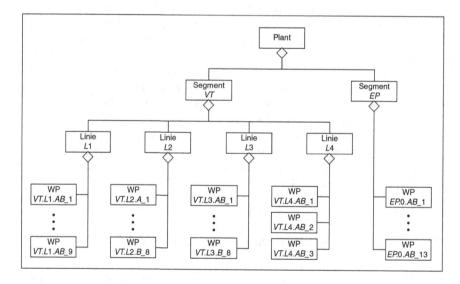

Fig. 4.10 Illustration of the investigated production system in the form of a UML object diagram

- **Lines** are denoted by *[Segment].L[x]*, where *[Segment]* is the segment in which the line (abbreviation "L") is found, and *[x]* is the running number of line within the segment.
- The **Workplaces** themselves are named *[Segment].L[x].[Product]_ [y]*. This denotes with *[Segment].L[x]*, the respective line in the corresponding segment, which belongs to the workplace. *[Product]* provides information on the product variant, whose production is supported by the workplace. Thus referring to the case study, the combination "A", "B" or "AB" is possible. *[y]* is used on the contrary as the sequential numbering of the workplaces within the parent sub-system. Accordingly *EP.0.AB_3* implies the third workplace, directly assigned to the segment "Endprodukte", without showing a line allocation. It contributes to product variant "A" as well as variant "B".

The hereditary principles provided for by the paradigm of object orientation allow the dynamic configuration and parameterization of the PSM. As a result, calculation parameters assigned in the reference model are easily transferred to system objects and, if necessary, expanded. In this way a *faster and less complicated configuration* of possible production models is attained, as was highlighted with the case study example during the flexibility analysis. Thus a slightly abstract replication of the factory was possible in ecoFLEX, which was true also for the alternative configurations required for the analysis (in accordance with **Requirement R3.4**).

The same is valid for the calculation functionality of Quantity, Mix and Expansion flexibility, which are also handed down and are then expanded with additional features. Because of the technical data *link of the flexibility metrics* with the PSM, it is possible within ecoFLEX to make separate reports for each system object covered, aside from the actual flexibility calculations. Examples of this may be seen in the issue of additional object-related characteristics for the better classification of the identified flexibilities (see field (2) in Fig. 4.4) or the existing possibilities for creating new, hitherto unconsidered system hierarchies (see Fig. 4.11) (corresponding to **Requirement R3.5**).

4.3.3 Verification of Flexibility Metrics

The claims made in Sect. 2.5.2 with the results presented in the case study analysis shows a very satisfactory fulfillment of the requirements for the flexibility metrics.

The specific *Mix-, Volume- and Expansion flexibilities* were able to be determined for all evaluation objects which were considered. By means of a simple procedure, conclusions could be made regarding the system's response to fluctuations in the quantity and product-/variant mix and also with respect to capacitive expansions. This is illustrated by the different flexibility assessments of the case study outlined in Sect. 4.2.3 (in accordance with **Requirement R2.1**).

The complete system does not consist of the object of observation alone, but also the subsystems summarized below, which are distributed across the observation levels provided for the hierarchical classification of the system, namely *workplace, line, segment* and *factory*. Flexibility deficits can therefore be easily assigned to the responsible system objects, which helps to limit the range of solution options quickly and accurately. This is also illustrated in Sect. 4.2.3. where the identification of individual flexibility vulnerabilities and the way in which to remove them is . addressed in detail (in accordance with **Requirement R2.2**).

A useful tool for the identification and control of the production flexibility deficits is the *quantifiable evaluation procedure* through which, depending on the flexibility type, a specific index for each detected object can be calculated. This ensures a quick and objective analysis of the user companies. In the absence of subjective influences, the construction of a multi-purpose workplace is recognized as the most favourable expansion alternative, although it initially appeared to be the worst due to its low maximum capacity (see Table 4.3). This remained unchanged even after multiple calculation loops with slightly modified parameter values of the expansion measures (in accordance with **Requirement R2.3**).

The *multi-dimensional flexibility character* of the developed evaluation methodology is also considered as requirement conforming. In each of the three flexibility evaluation techniques (see Sect. 4.2.3) the dimensions cost, time and variety are all equally considered. The evaluation methodology is applicable for both Multi-product and Single-product manufacturing and allows meaningful conclusions to be drawn on economical product creation. This is demonstrated by the different results of the case study example in Sect. 4.2.3, as the following representation of *Volume-, Mix- and Expansion flexibility* shows (in accordance with **Requirement R2.4**):

- The Volume flexibility of the analysed double-product factory amounted to 68.4% with the economic fluctuation in the output volume, which indicates the Variety dimension, lying between 289 and 917 units for the infotainment product variant *A*, and 354 and 1,121 units for the variant *B*. The determination of the respective break-even quantities and the maximum capacity was only possible while taking into account the *cost-* and *time dimension*. The concrete manifestation of such characteristics is directly related to the time-dependent production resources and the costs of their demands within a defined observation period.

- The *variety* dimension is reflected in the Mix flexibility by the average production gain deviation. It was shown using the case study that a large average deviation carries a high risk for economical production with a changing product mix composition. This was proven by the relatively low flexibility ratio of 43.1% of the factory, which due to changes to the existing ratio of the two product variants A and B, brings about greater profits as shown in the Table 4.4. Due to the economic background of these considerations, the *cost-* and the *time dimensions* are of significant importance since the determination of the flexibility index of 43.1% is directly dependant on the resource times and the corresponding resource costs of the factory.

- The Expansion flexibility also exhibits a multi-dimensional character. As with the Volume flexibility, the *Variety dimension* is also expressed by economic fluctuations in output levels. It is always based on that expansion measure which produces the best cost-benefit ratio and thus the least economic expenditure. For the example of line *VT.L2* this means production volumes between 4,867 parts (break-even quantity) and 23,641 units (target capacity), which corresponds to the expansion-based flexibility space "Alternative 3" (see Table 4.3). For this calculation, the inclusion of the *cost dimension* was essential since, amongst other reasons, the objective determination of a break-even quantity without cost information is not possible. The same is also true for the *time dimension,* which not only influences the resource times, but also sets the time period for the implementation of expansion measures. In the case study this was 6 months.

The question of the fulfilling the last, metric related requirement, the *comparability of the results* is not one without uncertainties. Indeed, as shown in the case study, the comparison of different expansion alternatives was possible with the help of the calculated flexibility indicators (see Table 4.3) and parameters could also be determined for a particular subsystem regardless of whether it was located on an equivalent system level in the hierarchy or not. Evidence of this was the identification and resolution of flexibility bottlenecks in the workplaces *EP.0.AB_3, VT.L2.A_7* and *VT. L2.B_8* (corresponding to **Requirement R2.5**). The comparability of flexibility indicators of systems from different sectors however, cannot be viewed as being concretely proven since the methodology was thus far only used in one branch of industry. Nevertheless, due to the quantifiable assessment approach which is based on economic parameters determined for each production system, it may be assumed that such comparisons are feasible. With the appropriate widening of the evaluation methodology, the proof of this statement should not be very difficult.

4.3.4 Verification of the Fundamental Applicability

The development of the evaluation method was shown to be in accordance with the requirements set out for it. This is demonstrated by the verification of ecoFLEX through the use of various defined test scenarios, as well as the verification of the production system model and the flexibility metrics based on experience from its applications in a production company. Subsequently, the general suitability of the methodology should be proven through the case study example. The basis for this was a specially formed "ecoFLEX work group," in which the previous application experience was discussed and assessed in terms of fulfillment of the requirements for applicability as described in Sect. 2.5.1. The group included 14 experts who were not necessarily from production companies, but also from various IT companies specializing in product and production data management and production virtualization.

To ensure that the assessment methodology is orientated to the mastering of practical flexibility problems, intensive talks with experts were held at various intervals during the design phase. Thus ensuring a development which conformed to the prescribed requirements. The consensus at the end of the "ecoFLEX work group" confirmed the *practical feasibility* of the methodology. The prerequisite for their appropriate implementation however, is the knowledge of the composition of the product mix and the sale prices of the products (see Table 3.1). Independent of this fact, the ecoFLEX software tool was presented at the Frankfurt Machine Tool Fair "EuroMold" and enjoyed great attention. Here many interested parties were convinced of the practical relevance of the evaluation methodology (corresponding to **Requirement R1.1**).

The un-ambiguous proof for the *cross-industry applicability* of the methodology could not yet be provided. That would require, as in the case study, a similarly detailed analysis of other production systems in various branches of industry. This opportunity has until now not presented itself to the author, reflecting in particular the lack of awareness of the evaluation possibilities offered by ecoFLEX. Nevertheless, it is based on the satisfaction of this requirement criterion that the development of the methodology took place completely dissociated from the case study example of the infotainment area which came later. Forming the important foundation for the concept of the evaluation method were the previous discussions with experts in the production and production-related IT field, from which the example production system presented in Sect. 6.4 emerged. It took into account the various types of special cases in real existing production systems and served as a test instrument in the realization of the concept. The results of this are flexibility ratings based on quantifiable parameters (see Table 3.1), which should be determinable for each production system. According to the working group participants, it is irrelevant whether the object being observed is a wood-, metal- or plastic processing plant or any other production system; or whether their outputs are measured in parts, litres or tons (corresponding to **Requirement R1.2**).

The consensus amongst experts involved in the working group was that the fulfilment of the requirements can also be expected in the *data collection*. Thus satisfying the demand for *completeness*, which is also a prerequisite for the carrying out of the flexibility calculations by means of clearly defined cost-related and non-cost-related parameters. They are considered as variable elements that apply to each production system and whose data is assigned to the appropriate flexibility relevance. In contrast, data with little or no importance is discounted (see Sect. 3.2.1.1). The *simplicity* in data collection ensures an object-oriented reference model, which due to the inheritance principle allows the re-use of existing data or parameters, while safeguarding their information content. As a corroborating example of a requirement-conforming data collection, please refer to Table 4.6 below. It shows, based on the case study, the completely levied, cost- and non cost-related, flexibility relevant data values for the line *VT.L4*. According to the degree of abstraction, the data for the three workplaces allocated to the line are thereby also outlined here. The principle of inheritance allows these workplaces to be considered as independent system objects, which avoids further, redundant data capturing (in accordance with **Requirement R1.3**).

According to all the ecoFLEX task force participants, the need for a customised *structuring and detailing* of the data necessary for the evaluation methodology was satisfied. On the basis of the paradigm of object orientation, the risk of result

Table 4.6 Cost- and non cost- related calculation parameters from line VT.L4 of the "Verbauteile" segment of the case study

Non-cost-related parameters	
FP	Non
IP	*VT-A3_1; VT-B3_1; VT-A3_2; VT-B3_2; VT-A3; VT-B3*
M	$M = \{VT\text{-}A3\_1; VT\text{-}B3\_1; VT\text{-}A3\_2; VT\text{-}B3\_2; VT\text{-}A3; VT\text{-}B3\}$
W	$W = \{VT.L4.AB\_1; VT.L4.AB\_2; VT.L4.AB\_3\}$
R	$R = \{r_1, r_2, r_3, r_4, r_5, r_6\}$, whereas $r_1 = \{VT.L4.AB\_1; VT\text{-}A3\_1\}, r_2 = \{VT.L4.AB\_1; VT\text{-}B3\_1\},$ $r_3 = \{VT.L4.AB\_2; VT\text{-}A3\_2\}, r_4 = \{VT.L4.AB\_2; VT\text{-}B3\_2\},$ $r_5 = \{VT.L4.AB\_3; VT\text{-}A3\}, r_6 = \{VT.L4.AB\_3; VT\text{-}B3\}$
WHM	$WHM = \{WHM_1, WHM_2, WHM_3, WHM_4, WHM_5, WHM_6\},$
t_{max} *(WHM)*	$WHM_1 = 144{,}000s; WHM_2 = 180{,}000s; WHM_3 = 208{,}800s; WHM_4 = 288{,}000s;$ $WHM_5 = 345{,}600s; WHM_6 = 403{,}200s$
$t_{PT}(r_k)$	$r_1 = 50\ s; r_2 = 50\ s; r_3 = 18\ s; r_4 = 18\ s; r_5 = 140\ s; r_6 = 140\ s$
$t_{aIT}(WP)$	$t_{aIT}(VT.L4.AB\_1) = 10\% \cdot t_{max}(WHM); t_{aIT}(VT.L4.AB\_2) = 2\% \cdot t_{max}(WHM);$ $t_{aIT}(VT.L4.AB\_3) = 2\% \cdot t_{max}(WHM)$
$a(r_k)$	$r_1 = 3\%; r_2 = 3\%\ s; r_3 = 1\%; r_4 = 1\%; r_5 = 1\%; r_6 = 1\%$
s(MI)	No sale prices, because *VT-A3_1; VT-B3_1; VT-A3_2; VT-B3_2; VT-A3* und *VT-B3* are non sellable *IP*
Cost-related parameters	
$C_{var}(r_k)$	$r_1 = €1.05; r_2 = €1; r_3 = 0.85\ s; r_4 = €0.85; r_5 = 1.90\ s; r_6 = 140\ s$
C_{Fix}	$C_{Fix}(VT.L4.AB\_1) = €644.9; C_{Fix}(VT.L4.AB\_1) = €510.87;$ $C_{Fix}(VT.L4.AB\_1) = €538.92; C_{Fix}(VT.L4.AB\_1) = €1{,}442.31$

distortion may be encountered through multiple overlapping or the neglect of different circumstances. All cost- and non cost-related calculation parameters are recorded as so-called attributes in the main class "Production System" with their assigned data values which are passed on to the four sub-classes "Factory", "Segment," "Line " and "Workplace". This avoids redundant and possibly conflicting value assignments and also assigns only the relevant flexibility information to a system object. It also refers to the various metrics for calculating flexibility as well as the resulting figures (see Fig. 3.13). In addition the cost accounting framework is also applied, due to the brisance of a standard and easily reproducible cost structuring during the assignment of cost-related parameters. It sets out the framework for the complete cost structure adjusted to the degree of detail (level of observation) (in accordance with **Requirement R1.4**).

A further advantage related to the object-oriented approach is the reusability and abstraction ability of the different evaluation objects in a production system. As a result, the collected flexibility information can be combined with other objects or even in new objects that deviate from the observation hierarchy. Proof of this was demonstrated in the "ecoFLEX task forces" with reference to the practical example, the creation of a new object called "manufacturing cell", which was assigned the lines and *VT.L1* and *VT.L2* and their workplaces. The associated implementation effort for a programmer incorporated in the source code, regardless of the GUI, was only 6 min to view the new calculation results for the factory using the "Eclipse development environment" (see Fig. 4.11). GUI adjustments on the other hand, require an additional maintenance effort as for example in the integration of a new symbol for manufacturing cells and their associated Wizard functionalities, which takes about 3 h. According to experts, it is based upon this fact that the demand for a *dynamic, case-specific customization* of the evaluation methodology is accommodated (in accordance with **Requirement R1.5**).

The last requirement found in the task force, concerned the assessment of the *reproducibility and transparency of the calculation results* for flexibility. There were differing views on the representations of flexibility parameters as percentiles. The base value on which the figures are set in proportion, is not necessarily visible. As an alternative, it was discussed that percentage multiplication be completely waived and instead to indicate the calculated value directly which would lead to, for example, a flexibility index of 95% having the numerical value of 0.95. This solution was accepted to a certain degree because it also preserved the ability of the flexibility to be assessed however, the majority argued for the representation as a percentage, because it has a stronger association to the underlying views on flexibility given in this book (see Sect. 2.2.3).This is how quantity fluctuations are evaluated in relation to their economic reference limits, changes in product mix in terms of average profit losses and expansion provisions related to their specific cost-benefit ratios. Overall, the experts were convinced by the existing valuation approach which is built of quantified, logically successive steps. This lead to a generally positive feedback since it eliminates the influence of subjective factors. The unanimous opinion was that all flexibility parameters could be understood and checked for accuracy. "Unintentional" evidence of this is provided by the

Node	Volume flex	Break even	Ratio capacity	Mix flexibility	Expansion flex
Fabrik:	68.41 %	28513	31596	43.1 %	0.0 %
Verbauteile (VT):	68.41 %	20134	22311	37.4 %	0.0 %
Cell	68.41 %	13092	14507	37.4 %	0.0 %
Line 1 (VT.L1):	79.30 %	8243	12800	0.0 %	0.0 %
VT.L1.AB_1:	94.87 %	984	6720	0.0 %	0.0 %
VT.L1.AB_2:	89.95 %	964	3085	0.0 %	0.0 %
VT.L1.AB_3:	87.47 %	935	2613	0.0 %	0.0 %
VT.L1.AB_4:	95.13 %	935	6720	0.0 %	0.0 %
VT.L1.AB_5:	79.3 %	917	1424	0.0 %	0.0 %
VT.L1.AB_6:	92.06 %	889	3919	0.0 %	0.0 %
VT.L1.AB_7:	98.01 %	889	14399	0.0 %	0.0 %
VT.L1.AB_8:	91.27 %	862	3458	0.0 %	0.0 %
VT.L1.AB_9:	90.41 %	862	2892	0.0 %	0.0 %
Line 2 (VT.L2):	68.41 %	4848	5372	54.2 %	0.0 %
VT.L2.A_1:	93.27%	376	1960	0.0 %	0.0 %
VT.L2.B_1:	90.29 %	460	1660	0.0 %	0.0 %
VT.L2.AB_3:	83.18 %	837	1599	54.2 %	0.0 %
VT.L2.AB_4:	89.60 %	811	2733	54.2 %	0.0 %
VT.L2.AB_5:	91.09 %	787	2842	54.2 %	0.0 %
VT.L2.AB_6:	92.51 %	787	3380	54.2 %	0.0 %
VT.L2.A_7:	74.16 %	354	480	0.0 %	0.0 %
VT.L2.B_8:	68.41 %	433	480	0.0 %	0.0 %

Fig. 4.11 Screenshot of the calculation results of ad hoc incorporated manufacturing cell displayed in the Eclipse development environment

case study, where the maximum output levels of the workplace EP.0.AB_3 (see Sect. 4.2.3) calculated with ecoFLEX deviated significantly from the values in the factory simulation of the user company. A brief review of the time conditions (see Formula 3.4) immediately excluded a calculation error on the part ecoFLEX. The subsequent inspection of the digital factory planning tool clearly indicated conflicting input values regarding the processing times. It was highlighted that the error was an illogical connection between the manual and automatic product processing times of the said workplace. After its correction, the maximum output volumes between ecoFLEX and the digital factory planning tool were once again consistent (in accordance with **Requirement R1.6**).

Chapter 5
Summary and Outlook

The requirements for production systems are constantly changing as a result of changing competitive conditions and the associated performance targets regarding time, quality, cost and innovation. The ever-increasing planning uncertainties as to type (product/variant mix) and extent (amount) of products to be manufactured pose a difficult task for production companies and lead to a growing demand for flexibility. In this context, flexibility assessment methods of production systems play a significant role in allowing meaningful conclusions to be made regarding existing technical and organizational scope of action, which allows the creation of an optimized level of flexibility. Although manufacturing companies have offered a number of opportunities through this and multiple research activities in this area were, and continue to be, made, there is still no visible acceptance of such assessment tools in an industrial environment. The main reasons for this lie in the difficulty to conform to the multidimensional nature of the flexibility and to simultaneously allow consistent, focused appraisals for the different areas of a production system. But it is precisely the lack of established methods for flexibility assessment in corporate practice, despite the digital factory planning tools used, that leads to the building of suboptimal production infrastructure over and over again. The results are often considerable flexibility deficits that can endanger the cost-effective manufacture of the products during "turbulent" times, which are amplified by the current financial crisis.

5.1 Summary

This book presents a significant contribution to solving the above-mentioned problem for flexibility assessment of production systems. The core elements are the three evaluation methods for the assessment of Volume-, Mix- and Expansion flexibility, the object-oriented production system model as well as the ecoFLEX implementation. The starting point for their development is the **guiding research question** formulated in Sect. 1.3, which narrow the field of observation and lead to the cognition process in this book. The important results of this process have been summarized below.

S. Rogalski, *Flexibility Measurement in Production Systems*,
DOI 10.1007/978-3-642-18117-7_5, © Springer-Verlag Berlin Heidelberg 2011

Several meanings, influencing factors and -objects for production systems can be extracted from the literature. An *analysis* of the evaluation method focused on the objective of the book brought the following findings to light:

- Production systems are goal-related summaries of the resources: equipment, supplies, personnel and material and characterize themselves through organizational, technical and economic characteristics.
- Their functions include not only the technical production process but also the planning, controlling and maintenance of the production process.
- Depending on the demand for explanation of the production system, their detailing can be done on different observation levels, each representing a specific set of system objects which have basic, common and level-specific characteristics. For a useful distinction between these levels, the hierarchical sub-divisions called factory, segment, line, and workplace are used.
- The various objects in a production system are controlled by multiple links, whose spatial-temporal allocation and hierarchical classification describe the system structure. The structure involves certain degrees of freedom which vary in their distinction for individual system objects, depending on external and internal changes.
- All processes running in a production system are either directly or indirectly related to the actual services rendered, whose results are material goods, called the product.

Based on these characteristics of production systems, the term "Production system" can be defined as follows:

A production system is an allocation, aligned with physical value-creation, of the resources equipment, personnel and material that are grouped together at various system levels for specific objects. These so-called system objects have the appropriate degrees of freedom, from which emerges a specific system dynamic for the response to external and internal changes.

The analysis for detecting the importance of flexibility in production systems showed that many opinions exist among experts as a result of the numerous inconsistent terminologies. As a consequence, there is a lack of common understanding of the scope and limitations of flexibility in the related concepts such as versatility, agility and adaptability. In principle however, the existing freedom of action or freedom of decision in production systems, which act upon the change, are denoted as flexibility. The following three dimensions are characteristic of their description:

- Variety
- Cost
- Time

The fundamental understanding of manufacturing flexibility developed by the author in the course of his research can be summarized in the following definition:

The flexibility of a production system describes its technical and organizational freedom of action to react to environmental uncertainties arising from economically justifiable adjustments or changes to the system structure and resources; so that the estimated production targets are met. The concrete Flexibility measure of a system can be determined by the dimensions variety, cost and time.

In addition to this general approach, the application context in which the production systems have to respond flexibly must also be included. The corresponding literature distinguishes between different types of flexibility, which vary however, as a consequence of the diverse classification possibilities as well as differing terminologies. Taking into account the identified, current developments and challenges in industrial practice (see Sect. 1.2) the range of flexibility types is limited for this book to *Volume- Mix- and Expansion flexibility*. Their impact and definitions have been clarified in Sect. 2.2.3. Based on the overall findings for the flexibility of production systems, the following criteria for a practical flexibility recording are derived:

- Evaluation ability at the observation levels factory, segment, line, and workplace
- Consideration of the dimensions time, cost and variety
- Cross-industry applicability, as well as comparable and objectively determinable evaluation results
- Evaluation of flexibility in terms of responsiveness to volume fluctuations, to changes in the Product-/Variant mix and on expanding capacity

Extensive literature searches show that none of the existing flexibility assessment methods can meet these demands to a satisfactory degree. Hence the corresponding need for research. This is subject to certain *requirements* (see Sect. 2.5), which are essential for the development of a flexibility evaluation methodology which conforms to the target requirements. Due to the similar importance of these requirements, they can be categorized into *basic usability, flexibility metrics, production system model* and *software implementation*.

The requirement definition made in this book sets out the principal direction towards the ***development of evaluation methods***. The basic idea touches on quantifiable dimensions used to make Volume-, Mix- and Expansion flexibility, including their subsystems, measurable. The individual flexibility types are assigned the following *quantifiable dimensions*:

- **Volume flexibility:** Break-even point, cost-efficient maximum capacity
- **Mix flexibility:** system optimal profit, product-specific profit deviation
- **Expansion flexibility:** alternative-specific break-even point, target capacity

All calculations in this context to determine flexibility indices will be combined under the concepts of flexibility evaluation method or flexibility metrics. Since the different dimensions to be quantified often exhibit varying temporal as well as cost- and product related restrictions, their boundary values (minima or maxima) are to be determined. These can be traced back to optimal, restriction conforming production plans. Such a plan describes the best possible utilisation of a system's resources regarding their type and complexity for specified period and number of products. The calculation of these production plans lead to so-called optimization problems that in principle can be considered to be linear. In order to solve these problems, the simplex algorithm is used. The prerequisite for this is the setting up of a mathematical model, which provides a distinction between three types of calculation parameters:

- Non cost-related calculation parameters (see Table 3.2)
- Cost-calculation parameters (see Table 3.3)
- User-dependent calculated parameters (see Table 3.4)

Because these parameters are counted as variable elements that are assigned different values depending on the operational systems in production, it needs to be clearly defined which criteria are to be used in the approach. This especially concerns the cost-related calculation parameters whose value assignments are closely related to the operational cost and performance and which have a direct impact on the quality of flexibility calculation results. It is therefore in the interests of a uniform, complete cost accounting (one which is based on the observation levels of production systems) to orientate towards a *cost calculation reference* approach that is modelled on part costing procedures with differentiated cost treatment (see Table 3.18).

In order for flexibilities for selected objects in a production system to be quickly calculated and to also remain easy to understand for the user, both the calculation parameters and the flexibility evaluation model are linked to the yet to be tested, real-world analysis object/production system via a so-called production system model. This model is seen to be an abstract representation of evaluation-relevant system objects in a neutral notation, so that their flexibility-related dependencies are identified and flexibility deficits easily assigned to the responsible positions. It follows the paradigm of object orientation and thereby allows easy, dynamic model configurations of the structure of the production system to be evaluated.

The practicality of the evaluation methodology developed in this book can be confirmed through a comprehensive **verification** in industrial application. This is done based on the production system of a series producer of infotainment systems, where the application experiences made with the involvement of expert opinions with the requirements of Sect. 2.5 were evaluated. The basis of verification is a specially developed software tool called ecoFLEX, which implements the mechanisms of the evaluation methodology. Through this it shows the requirement-conforming applicability through which allows the quantification of flexibility of the production systems and the assignment of individual system objects. The analysis possibilities applied in practice, such as the quick and easy identification of flexibility-related vulnerabilities and comparability of solution alternatives, highlight the user benefits of an ecoFLEX supported flexibility investigation and demonstrate the successful implementation of the evaluation methodology.

This can be further demonstrated by practical experience with ecoFLEX not included within this book, gained in other large companies outside of the infotainment industry and also in various medium sized production companies. There too, both unexpected and remarkable deficits in the production and change management came to light. The potential from the multiple applications for ecoFLEX are clarified by the following examples [Roga-10]:

- **Shortening of the decision making in investment projects:** the selection decision for the purchase of a new automated workplace at the production site of a medium sized producer of stamping and assembly technology with about 140 employees, called for an average cumulative input of six person-months (PM), which incur an average cost of approximately €54,000. Through the ecoFLEX solution, due to a significantly improved transparency of the causal dependencies in the production, this expense could be to reduced by up to 70%, which corresponds to a cost saving of €37,800 for each procured automated workplace. Due to regular product changes at the site, an average of four new job positions a year are currently required. This gives the ecoFLEX solution a total saving of €75,600 per year.

- **Reduction of follow-up costs through significantly improved selection decisions:** based on uniform, objective evaluation fundamentals, a consistent and transparent flexibility analysis was attained that easily identified flexibility deficiencies in production systems easily and correctly assigned and eliminated them. As a result, the follow-up costs can be reduced when purchasing new equipment, since the increasing the predictability is improved. In this way it was possible for the aforementioned medium sized company to reduce the likelihood of avoidable follow-up costs from 33.3% to 10% by reaching a sub-optimal investment decision. This results in average savings of €15,000 with each newly acquired workplace, giving a total saving of €30,000 a year.

- **Risk minimization for factory planning:** Regardless of the digital factory planning tools used by different user companies, weaknesses were found in the design and dimensioning of the production equipment, based primarily on their economic effects. The utilisation of ecoFLEX allowed a significantly improved factory planning that is tested for cost-effectiveness and which provides the right degree of flexibility. This allows the representation of economically viable alternative configurations of different production systems, which minimize its dependence on certain products or product variants, which ensures both an improved medium-term and long-term protection of cost-effective production.

- **Improved planning of personnel and the production program:** In addition to a pure flexibility analysis, additional parameters such as break-even levels or time- and cost-optimal production programs are determinable through the use of ecoFLEX. This also enables the evaluation of, for example, ad-hoc contracts in a current production program, so that they take into account cost and set-up configurations as well as allowing lead and holding times to be processed efficiently. In addition, the most cost efficient staff utilization for the processing of pending production orders can be determined, to which the most profitable work time is allocated. These are functions that are not covered by existing PPS and ERP-Systems. For the user companies, the use of ecoFLEX resulted in valuable cost savings in terms of a short to medium term safeguarding of their economical production.

As a result of such applications of ecoFLEX, it is particularly possible for small and medium enterprises to keep up with their competitors (including those from low-wage countries), despite the ever-increasing competition with tight cost and time budgets and an increase in complex production relationships. This is due to, on the one hand, improved on-time delivery and the resulting customer retention, which ensures a steady demand and promises future sales. On the other hand, the unique evaluation capabilities of the developed methodology, cause a targeted and dynamic approach to the production resources personnel, material and equipment. This avoids unnecessary additional costs for the inefficient use of resources or ineffective adjustments to the production infrastructure, such as construction and decommissioning of production equipment and creates funding for future investments [Roga-10]. Thus, secure existing jobs and creating more, for example, under the particular conditions of the economic and financial crisis by the acquisition or the establishment of additional production equipment that are integrated into the existing network and thus strengthening the company's personnel. The financial security and gains resulting from this benefit not only employers and employees in the production area, but also the national economy. Through stable or rising profits and wages the consumer economy is also strengthened, which has a positive effect on downstream businesses and industries. The resulting additional income taxes allow a strengthening of social security systems.

5.2   Outlook

As a result of the existing planning uncertainty in terms of capacitive demand fluctuations, changes in the product-/variant mix and capacitive expansion requirements, frequent adjustments and changes to the current production needs are inevitable. For this reason, companies increasingly find themselves confronted with the problem of a suitable recording of economic freedom of their production systems. The experience made in this area with the software ecoFLEX in the practical field of application, is testament to the usefulness of the developed evaluation methodology. As shown in Sect. 4.2, flexibility deficits can be identified using ecoFLEX by the quantification of flexibility margins within production systems and can be subsequently resolved. As a consequence, planning security and response time in recognizing and implementing flexibility-increasing measures are improved. In addition, costs attributed to flexibility weaknesses are avoided. Overall, the methodology supports a short, medium and long term protection of economic production. Linked to this, with the goal of compliant user benefits (see Sect. 1.3), is a crucial precondition. It refers to continuous and consistent data in each user company because the quality of the evaluation results and the associated success always depends on the quality of input data.

Nevertheless, the efficiency of the presented method of evaluation continues to grow. One approach would be the expansion of a special algorithm that recognizes the flexibility deficits in the production system and based on that offers *automatically generated alternative solutions* that allow the optimal configuration of volume, mix and Expansion flexibility. This could only be done based on the known system objects, whose object-specific parameter values are a prerequisite for the alternative determination. Thus, for example, an additional workplace proposed in this way, would be an exact copy of an already existing one. Additionally, the integration of a knowledge database would also be conceivable, which contains information on production equipment of different manufacturers, thus improving the quality of the solutions. Although the application of the algorithm does not imply that the decision on the measures of action is removed entirely from the user, it does however simplify the work of searching and selecting suitable alternatives.

Through this approach, it is valid in future to include the *network level* in the flexibility investigation. The reason is the frequently encountered product- and resource-related interdependence of manufacturing companies in the form of temporary production networks. Their goal is to compensate the high planning uncertainty induced by the turbulent environment on inter-company levels through the appropriate configuration of the network. [Milb-02] [Petr-06] [KRS-06]. In this context, reliable statements about the supply chain in procurement options of raw materials, purchased parts, etc. take on a crucial role. It is therefore recommended in addition to the admittance of the network level in the evaluation methodology, also a completion of the already considered flexibility types for the *procurement flexibility*. This offers opportunities for the assessment of different procurement approaches and their success.

In the interest of a universal and established measure of manufacturing flexibility, *certification* is also an aim of this work. Similar to the ISO 9000 certification,[1] requirements for the fulfilment of a certain degree of flexibility are defined, which are informative for the company- internal implementation and also serve as confirmation of certain standards for third parties. For example, production planners who deal with the organization of complex networks of production, would have an improved decision-making basis, to decide on the incorporation of a new production company in an existing network. Environmental turbulence can thus be offset by an optimal configuration of the targeted production network. However, requirements for this are flexibility evaluation methods that are applicable across industries and deliver comparable and reproducible flexibility values, as represented in the methodology developed here.

Finally, it is important to note that the concept presented in this book allows for a new, holistic way of viewing and evaluating flexibility of production systems reached at an abstract level. Beyond the directly related, scientific contribution, a comparability of production systems (even from different industries) is achieved through the application of the presented evaluation methodology. This offers the possibility for future research to include flexibility as a measurable quantity for the requirement-conforming selection and design of production systems.

---

[1]The ISO 9000 series of standards includes principles for the provisions for quality management. The norms falling into this category form a common and coherent set of standards for quality management systems [Hans-06].

Chapter 6
Appendices

6.1 Basic Knowledge

The purpose of this chapter is to clarify important background knowledge about business cost accounting, the character of the Simplex-algorithm as well as paradigms of object orientation. They present a necessary condition for understanding the procedure of the development of the evaluation methodology in Chap. 3.

6.1.1 Methods of Cost Accounting

In Sect. 3.3 the cost accounting reference frame is defined. This accounting allows to gather and to attribute costs and the sales prices that arise during the internal value-added-process as part of the internal accountancy [Eber-04] [Götz-04]. In the following, their main characteristics and differences will be explained.

6.1.1.1 Classification Criteria of the Cost Accounting System

Basically, the cost accounting system can be classified by two criteria. The first criterion is the time-reference, the second is the volume of included costs (see Table 6.1) [Eber-04] [Götz-04].

According to the time-reference, cost accounting is differentiated in the *actual-, normal cost accounting and planned cost accounting*. The former comprises the operational use of resources of an accounting period on the basis of the costs that have actually arose, the so-called actual costs. As this occurs retroactively, i.e., after the technological value added, this accounting is focused on the past. The normal cost accounting is a similar procedure, however, the accounting is carried out using mean actual costs of past periods. This is advantageous as random value fluctuations are of no great consequence. In contrast to the other two accounting procedures, the planned cost accounting is future-orientated, so that it is suitable for plan/actual comparison. Its advantage is the inclusion of the cost accounting in the

S. Rogalski, *Flexibility Measurement in Production Systems*,
DOI 10.1007/978-3-642-18117-7_6, © Springer-Verlag Berlin Heidelberg 2011

Table 6.1 Classification of the cost accounting systems according to [Burd-02]

		Time reference		
		Current cost accounting	Normal cost accounting	Planned cost accounting
Amount of attribution	**Full cost accounting**	Current cost accounting on the basis of full costs	Normal cost accounting on the basis of full costs	Planned cost accounting on the basis of full costs
	Direct cost accounting	Current cost accounting on the basis of direct costs	Normal cost accounting on the basis of direct costs	Planned cost accounting on the basis of direct costs

operational planning process and thus the given control of the economy. Those costs that have to be calculated are to be considered as norm-parameters. That is why we speak of standard costs, budgeted costs, estimated costs or target costs [Eber-04] [Götz-04] [Olfe-05].

According to the amount of attribution of the cost accounting systems, the second classification criterion differentiates between direct and full cost accounting. The full cost accounting system distributes full costs as detailed as possible (appropriate to the originator) to the appropriate reference object, the so-called cost object. In contrast, direct cost accounting contributes only certain parts of the total costs to the cost object. Only performance-related variable costs and relevant fixed costs are regarded, whereas other irrelevant fixed costs remain unconsidered [Burd-02] [Götz-04]. Table 6.1 summarizes the classification criteria of the cost accounting system.

A brief overview of relevant aspects of full- and direct cost accounting will be given in the following.

6.1.1.2 Full Cost Accounting

The basic concept of full cost accounting is to differentiate between *direct costs* and *operating costs*. Every cost type has to be checked with regard to what extent it can be directly attributed to a cost object (direct costs), or to what extent it is distributed to different cost objects (operating costs). The former applies if a direct cause-relation exists between the technological value added and the use of resources, for example: production materials, packing and tool costs or piecework pay. However, operating costs apply for several goods and services in a company at the same time and thus cannot be directly related to one type of goods or service. Examples are costs for buildings, additives, operation supply items or pay for the production management [Eber-04] [Götz-04] [Olfe-05].

Full cost accounting differentiates between *cost type accounting, cost centre accounting* and *cost object accounting*, whose relation is demonstrated by Fig. 6.1. In the following these three kinds of calculation will be explained in detail.

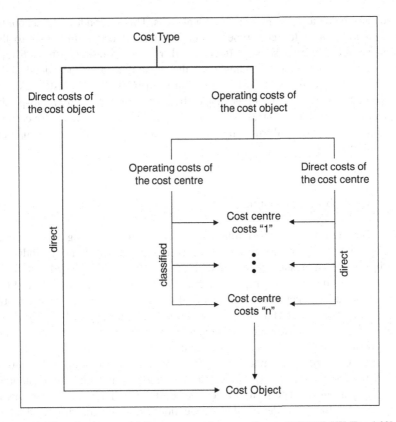

Fig. 6.1 Cost flow that is part of full cost accounting according to BURDELSKI [Burd-02]

Level 1: Cost Type Accounting

Cost type accounting is the basis for the entire cost accounting as it gives information on the cost structure and the cost level of a company [Eber-04]. In this connection the entire use of resources in a period, which is caused by the operational technological value added, is systematically gathered and evaluated. It can often be classified as personnel-, material-, energy- and maintenance costs, as well as taxes, fees, insurances, other or calculative costs [Seic-01]. The preparation of costs that results from the cost type accounting enables an optimal execution of the cost centre accounting and cost object accounting [Eber-04].

Level 2: Cost Centre Accounting

Cost centres are places where costs arise or are attributed and are considered to be operational sub-areas whose monetary accounting has to be carried out independently. They are classified by certain criteria as for example activities, causation components, calculative factors or functional areas. This second level of accounting occurs subsequent to the cost type accounting and relates arising costs to their particular origin. Its purpose is to allocate work-related operating costs to different cost centres, to control the efficiency of the technological value added and to plan operating costs for each cost centre [Burd-02] [Götz-04] [PlRe-06]. Another purpose that is very important is distributing operating costs to multi-product companies. There, cost centres have to be appointed according to operational conditions that have to conform to the production program and to the design/course of action [Coen-03].

Level 3: Cost Object Accounting

In the third step of cost accounting, direct costs and operating costs that are determined indirectly by the cost centre accounting are distributed to individual cost objects. Thus, a short-dated income statement can be carried out for an accounting period and for each product group or an entire company. Cost objects are those operational benefits (from received goods and services) per accounting period that use resources. Thus, they "bear" the costs. Examples are benefits of sales (as for example products, orders), storage activities (as for example positive stock balance difference of products and intermediate products) or in-company benefits (as for example self-made machines). The cost object accounting differentiates between two domains, the cost object accounting per unit and the cost object accounting per period. The former calculates costs of individual unit sizes, as for example unit-, lot- or order related production costs. In contrast, cost object accounting per period determines costs that are needed to determine earnings before interest and taxes (EBIT) for a relatively short accounting period, as a general rule [Burd-02] [Götz-04].

6.1.1.3   Direct Cost Accounting

In contrast to full cost accounting, direct cost accounting does not allocate fixed costs to single technological values added in as detailed a way as possible. This avoids inaccuracies and sources of errors that come from the apportion- and attribution of costs. Instead, this accounting procedure is orientated on "the user pays principle". Consequently, those costs that can be attributed directly to a reference object (direct costs) have to be captured, while fixed costs remain as a block [VDI-95] [Burd-02] [Götz-04].

Basically direct cost accounting is organized into two basic designs, in systems based on direct costs and systems based on variable costs. The latter is again divided, with the costs originating either from global or differentiated considerations of fixed costs. The commonness that each of these system features is the

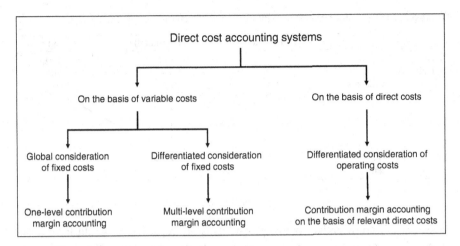

Fig. 6.2 Classification of the direct cost accounting systems

accounting of so-called contribution margins[1] [Kich-98] [Burd-02] [Kale-04]. Figure 6.2 demonstrates the hierarchic classification of the direct cost accounting systems which will be presented in short in the following.

Direct Cost Accounting System Based on Variable Costs with Global Consideration of Fixed Costs

This kind of cost accounting is a single-stage procedure of the contribution margin accounting whose total costs consist of fixed and variable costs. Here, as a general rule, direct costs of relative cost objects and their operating costs that can be attributed directly are to be considered as variable, whereas fixed costs account for the remaining part of operating costs. When the earnings before interest and taxes (EBIT) of a company are calculated, variable costs first have to be related to cost objects. The contribution margin results from subtracting variable costs from the product-specific turnover. After that the remaining fixed costs are attributed to the entire company as a block (see Table 6.2) [Burd-02] [Kale-04] [Götz-04].

When it comes to decision-making, this kind of direct cost accounting is very important in regard to assortment- and product planning in a multi-product company. The reason why certain products are selected to be part of a product group in a range of production is their contribution margin, rather than positive earnings before interest and taxes (EBIT). A positive contribution margin can only contribute to reducing the amount of fixed costs, even if the earnings before interest and

---

[1]The contribution margin is that part of revenues of sales which, after having subtracted fixed costs, remains for the production of a product and can thus be used for covering fixed costs. As a consequence, the contribution margin provides information about the contribution that a product makes to the coverage of fixed costs or to the overall result of a company [Kale-04] [KeBo-98].

Table 6.2 Example of direct cost accounting with global consideration of fixed costs

	Product A	Product B	Product C
Turnover	900	600	1,500
Variable costs	−500	−300	−800
= **Contribution margin**	**400**	**300**	**700**

Sum of contribution margin	1,400
Block of fixed costs	−1,000
= **Operating result**	**400**

taxes (EBIT) are negative. If this product was not regarded, it would have been impossible to decrease existing fixed costs, so the earnings before interest and taxes (EBIT) would have been less than necessary [Scha-93] [Burd-02].

Direct Cost Accounting System on the Basis of Variable Costs with Differentiated Consideration of Fixed Costs

As well as the system of global considerations of fixed costs, the direct cost accounting system separates variable and fixed costs. However, fixed costs are allocated in steps to different reference objects according to a funds-purpose-relation. These steps can be classified in types of products, groups of products, cost centres, sector and companies. Fixed costs are subtracted from each of these reference objects. Thus, several steps of contribution margins are determined. This results in a hierarchy of accounting that is directed to that system, that features the highest fixed costs, so that only a small amount of fixed costs has to be attributed to the following higher step only [FHPP-99] [Seic-01] [Burd-02]. Table 6.3 demonstrates an example for determining earnings before interest and taxes (EBIT) on the basis of the differentiated considerations of fixed costs.

The system of this step accounting of the fixed-charge coverage accounting is a full cost accounting without any artificial ratio-based fixed cost attribution.[2] Its advantage lies in the fact that fixed costs can be attributed on every hierarchical level without any need of apportion. Furthermore, it can contain additional important information about the actual period result of single reference objects next to the contribution margins [FHPP-99] [Seic-01] [Burd-02].

---

[2]Ratio-based fixed costs attribution describes the allocation of fixed costs to the production volume in order to determine total costs per piece in a product unit [MHE-03].

Table 6.3 Example for direct cost accounting with differentiated consideration of fixed costs

	Product 1	Product 2	Product 3	Product 4	Product 5	Product 6
Turnover	900	600	1,500	1,000	2,200	3,600
Variable costs	−500	−300	−800	−800	−1,400	−2,300
=Contribution margin 1	400	300	700	200	800	1,300
Fixed costs of product types	−50	−100	−200	−50	−150	
=Contribution margin 2	350	200	500	150	650	1,300
Fixed costs of product groups	−150		−100	−200		−300
=Contribution margin 3	400		400	600		1,000
Fixed costs of cost centre		−350		−250		−150
=Contribution margin 4		450		350		850
Fixed costs of division			−300			−350
=Contribution margin 5			500			500
Fixed costs of company			−600			
=Operating result			**400**			

Direct Cost Accounting System on the Basis of Relative Direct Costs

The procedure of relative direct cost accounting developed by RIEBEL is a combination of cost type-, cost centre- and cost object accounting. Its purpose is to allocate problematic operating costs of cost objects and cost centres. It considers all cost types that arose in a period and is, in this sense, similar to full cost accounting. However, it should not be equated with it, as RIEBEL does not refer the terms "direct costs and operating costs" to cost objects, but differentiates between objects of attribution and reference objects. Similar to direct cost accounting on the basis of variable costs with differentiated considerations of fixed costs, he carries out a successive attribution of incurred expenses in single steps, which leads to different contribution margins. However, contents of different accounting objects cannot be compared with those of the differentiated considerations of fixed costs in this procedure. Thus, the sales revenue (turnover) that was realized by sold products is reduced by value-added costs of sales, for example. This results in the so-called reduced Net Operating Profit After Taxes (NOPAT). In the next step, the product-related, value-added costs have to be subtracted from this in order to obtain single product contribution margins. Then these contribution margins are summarized to product groups which then are subtracted from product group-related, value-added costs. This is identical to direct cost accounting with differentiated considerations of fixed costs. Thus, particular contribution margins of product-groups are determined, whose sum represents the gross earnings before interest and taxes (EBIT). Furthermore, direct costs per period of

Table 6.4 Structure of the income statement as part of relative direct costs accounting according to RIEBEL [Rieb-94]

	Product group 1		Product group 2		Sum
	Product 1	Product 2	Product 3	Product 4	
Turnover	**10,000**	**80,000**	**60,000**	**50,000**	**2,90,000**
Value-added costs of sales	−3,000	−2,400	−1,800	−1,500	−8,700
NOPAT	**97,000**	**77,000**	**58,200**	**48,500**	**2,81,300**
Product-related, value-added costs	−28,500	−15,900	−4,600	−3,900	−52,900
Contribution margin of product	**68,500**	**61,700**	**53,600**	**44,600**	**2,28,400**
Product group-related, value-added costs	−22,500		−9,300		−31,800
Contribution margin of product group	**1,07,700**		**88,900**		**1,96,600**
EBIT	**1,96,600**				**1,96,600**
Direct costs of production division per period	−28,200				−28,200
Direct costs of production-supported cost centers per period	−6,200				−6,200
Contribution margin of all products in production per period	**1,62,200**				**1,62,200**
Direct costs per period of the administration and operation in the company	−19,600				−19,600
Liquidity related period result	**1,42,600**				**1,42,600**
Not explicit costs	−45,000				−45,000
Net operating result	**97,600**				**97,600**

production division and production-supported cost centres have to be subtracted from this result. The difference represents the contribution margin of all products in the production (by means of the direct costs per period). If direct costs per period of the administration and operation in the company are subtracted from the contribution margin, the result is the liquidity related period result. Subtracting the non-explicit costs finally leads to the "Net Operating Result" [Rieb-94].

Table 6.4 demonstrates the described structure of this income statement as part of the relative direct costs accounting.

6.1.2 The Simplex-Algorithm

Following from Sect. 3.2, the calculation of indices of Volume-, Mix-, and Expansion flexibility are mainly based on determining restriction-conforming, optimal production programs. The formulation of the corresponding optimization problems that are supposed to be linear is premised, which are solved by the help of the simplex-algorithm. The simplex-procedure is described in the following:

With the simplex-algorithm, a finite sequence of vertexes of the valid range M of L can be determined for the standard problem L of the linear optimization. These vertexes are so-called basic solutions of L, at which the value of the objective function decreases constantly at the transition from one vertex to another. In the case that L has a solution, the last vertex of this sequence is considered to be optimal (see Fig. 6.4, p. 158). In order to be able to apply the simplex-procedure, the standard problem L of the linear optimization first has to be adapted to a calculable form [NeMo-04] [WaSt-04] [DoDr-07]. Here the resulting optimization problem can be described according to Fig. 6.3.

From the given standard problem of the optimization, a simplex-table can be established (see Table 6.5). Above this table are the non-basic variables (NBV) that have the value 0 at the basis point. In contrast, those variables that feature a minus sign represent the basic variables (BV).

After having successfully established the simplex table, the simplex-algorithm can be applied to the optimization problem in steps as follows [NeMo-04] [WaSt-04] [DoDr-07]:

1. Minimization of the objective function; provided that a maximization problem exists, the equation has to be multiplied by -1
2. Enter the values in the simplex-table

Objective function	$Z \to \min$ $c \cdot x + \delta = Z \to \min$
Constraints	▪ $A \cdot x - b = -u$ ▪ $x \geq 0, u \geq 0$

Fig. 6.3 Structure of the optimization problem

Table 6.5 Abstracted simplex-table

NBV	x_1	x_2	$\cdots$	x_n	BV	
	a_{11}	a_{12}	$\cdots$	a_{1n}	$-b_1$	$-u_1$
	a_{21}	a_{22}	$\cdots$	a_{2n}	$-b_2$	$-u_2$
	$\vdots$	$\vdots$	$\cdots$	$\vdots$	$\vdots$	$\vdots$
	a_{m1}	a_{m2}	$\cdots$	a_{mn}	$-b_m$	$-u_m$
	c_1	c_2	$\cdots$	c_n	δ	Z

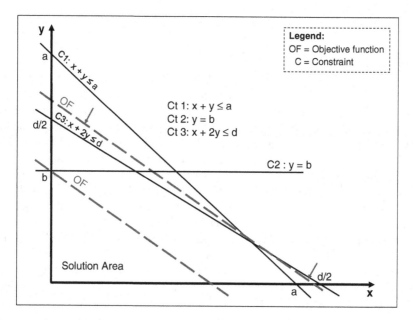

Fig. 6.4 Graphical demonstration of an example with three constraints

3. Finding the pivotal-column; the column with the smallest negative number c_n or if several equal numbers exist, the one with the smallest index n.

4. Finding the pivotal-row; the parts with the smallest quotient $\frac{b_p}{a_{pk0}} = \min\left\{\frac{b_p}{a_{pk0}}; \right.$ $\left. j \in G \text{ und } a_{jk0} > 0\right\}$.

5. Procedure of replacements; repeating till all $c_n \geq 0$ are:
   (i) Pivotal-element: $a \to \frac{1}{a}$
   (ii) Pivotal-row (without Pivotal-element): $b \to \frac{b}{a}$
   (iii) Pivotal-column (without Pivotal-element): $c \to \frac{c}{a}$
   (iv) Remaining elements: $d \to d - \frac{b \cdot c}{a}$
   (v) When all $c_n \geq 0$, the optimal solution is found; if not go back to step 3

where:

   $b \Leftrightarrow$ Row element of the pivotal-row with the same column index as d
   $c \Leftrightarrow$ Column element of the pivotal-column with the same line index as d

The Fig. 6.4 demonstrates the successive procedure of the simplex-algorithm in a graphical way.

6.1.3 Object Orientation

In the context of the fact that the production system model is a necessary part of the developed evaluation methodology (see Sect. 3.4), the principle of object orientation is of decisive importance. It provides a model- and data specific connection between the operational analysis object and the flexibility evaluation methodology. This results in the advantage of a relatively big, industry-independent freedom when production system models are constructed and parameterized. Hence, the effort of adaptations within the methodology is reduced, independently from the fact that they directly concern the flexibility metrics or the production system model itself. The paradigm of the object orientation will be explained in the following chapters.

6.1.3.1 The Object Orientated Approach

The term object orientation stands for a modelling approach that describes phenomena that are found in the real world. The advantage is the separation of static and dynamic aspects of an application area that needs to be developed. Certain properties are assigned to each object in this sector. An object has to be considered as an abstract model of a participant, that can complete orders, gives information about its status or changes it, and that is able to communicate with other objects [Stau-05]. According to their basic characteristics, objects have to be related to different classes as the following example demonstrates:

Example. The term "Automobile" summarizes different kinds of multitrack motor vehicles, that can be classified in *passenger car*, *buses*, *trucks* and *tractors* (see [KBA-09] also). They all share common characteristics like being motor-driven, are at least four-wheeled and transport persons or freight. However, passenger cars differ from buses in their construction and equipment. According to the German definition by law in "§ 4 clause 4 PBefG-Kraftfahrzeuge", these passanger cars are not suitable and not designed for carrying more than nine persons (driver included) [BMJ-09]. In contrast it is planned that a bus carries more than nine people and according to its construction and its equipment, a truck is constructed to carry goods. Tractors are characterized by pulling other, non-motorised vehicles (trailers, cars), that are supposed to transport goods or persons, for which they themselves are not constructed [KBA-09] [BMJ-09].

Besides these four categories of the "Automobile", it can be grouped in other subclasses. Trucks can be grouped into pickup trucks up to 3.5 tons (t), light rigid vehicles up to 7.5 t, medium rigid vehicles up to 12 t and trucks heavier than 12 t. Examples of light rigid vehicles are the "Mercedes-Benz 818 L Atego", the "MAN TGL 8.210" or the "Renault Medium 180 DCi". Due to an abstraction, all

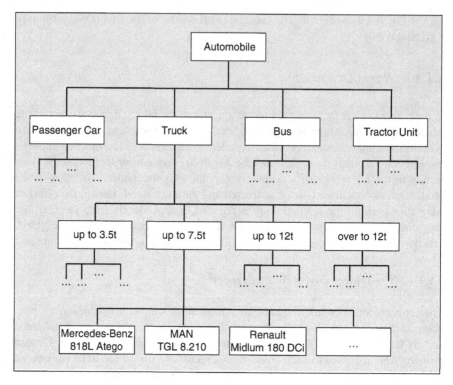

Fig. 6.5 Object orientation of the example "Automobile"

these trucks can be considered as "Automobiles" (see Fig. 6.5), whose common, general description is called "class" in the object orientation.

Similarly to the previous example, phenomena of nature or buildings that are created by humans can be classified in a multi-level hierarchy so that complex circumstances can be clearly arranged. For this the subsequent Fig. 6.5 gives a brief overview.

Object orientation has been strongly influenced by software development and results from rapid advancements in production engineering in the hardware sector that occurred in the early years of computer development. Due to the associated need of an accelerated software development/adaptation of the hardware, commercial software (mass) production was demanded, from which the methodology of object-orientated software development arose [Kees-98]. As a result of the object-orientated encapsulation, it was possible to maintain single modules of a program (or methodology) separately and independently from the overall structure. The first implementation of object orientation was performed with the computer language "Simula" in the mid-1990s. Today, the modeling language *Unified Modeling Language, UML*, is considered to be an established standard of the object-orientated programming [CSS-99].

6.1.3.2 Relevant Principles and Different Kinds of Representation in Object-Orientation

According to previous explanations, an object is in a certain state and reacts with a pre-determined behaviour upon information of its environment in the object-orientated development. Thus, it can be considered as an enclosed unity, which encapsulates specific properties and that communicates through a defined interface [Kees-98].

Each object is part of a superior class, which describes its construction plan or its data structure. This takes place by means of attributes as specific properties of a class that can be related to each other. Furthermore, a class features *methodologies* (functions) that cause the objects' properties. They describe class-specific tasks that are permanently available for the object, and that are usable if needed [CSS-99] [Balz-00]. Classes and objects are usually represented as a rectangle that is divided in three chapters, as demonstrated in Fig. 6.6 [Balz-00].

Analogous to Fig. 6.6, a class is defined as a summary of objects. However, there is no need to establish new attributes and methodologies when creating new classes. This can also be carried out by inheriting existing classes. Here, new classes (sub-classes) take over the characteristics of the original class (superior class) and expand these by determining additional attributes and methodologies next to the existing ones of the superior class. On the basis of Fig. 6.5 (p. 160), an example could be the expansion of the basis class "Automobile" with the attribute "tipping load" and the method "Tipping", which are considered in the subclass "trucks".

As long as the act of inheritance takes place in a series of several steps of a class hierarchy, it is called a so-called *inheritance hierarchy* (see Fig. 6.5, p. 160). It must not be marred by *multiple inheritance*, which allows a class to take over characteristics from more than one superior class. Thus, the class "Camp-mobile" could take over the characteristics of habitation and mobility of the two superior classes "House" and "Automobile" [CSS-99] [Dumk-00].

Figure 6.7 represents the two inheritance principles graphically in the UML-Notation.

In the case that classes communicate with each other or come into contact in any other way, this is expressed by an *association*. The demonstration of this circumstance is carried out by an association line, which simply provides information about the fact that objects of involved classes know each other. In general, their naming describes the direction of this connection and a more detailed specification can be carried out by characterisation of cardinalities and roles. These indicate the number of how many objects a certain object knows and how important it is (see Fig. 6.8) [Oest-06] [Balz-00].

A special form of association is *object composition* that represents a directed association between two classes and whose marking occurs by a rhombus at the end of the object composition. It contains "consist of" – semantics and provides information about how the whole thing is put together by single parts [Oest-06]. For instance, an automobile generally consists of a chassis, wheels, an engine and

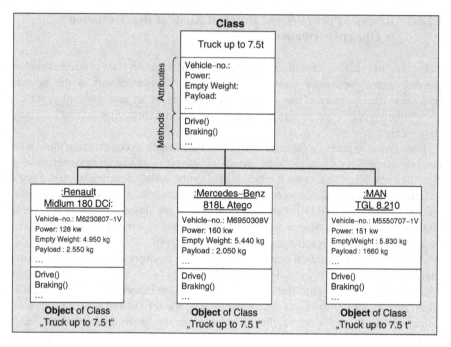

Fig. 6.6 "Light rigid vehicles up to 7.5 t" as an example to demonstrate the relationship between class and object

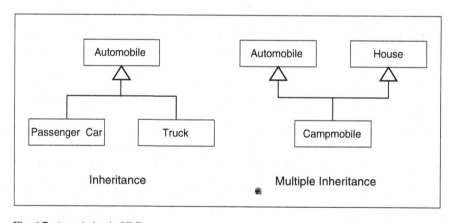

Fig. 6.7 Association in UML

the coachwork. However, in order to be able to strengthen this tie for an automobile and in order to emphasize the existence of these sub-objects, the end of the object composition is being marked with a coloured rhombus (see Fig. 6.9). This is called *function composition* [Oest-06].

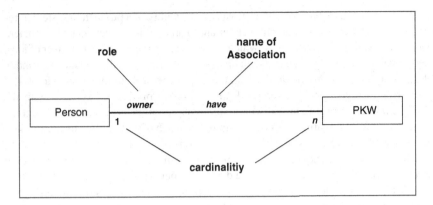

Fig. 6.8 Association in UML

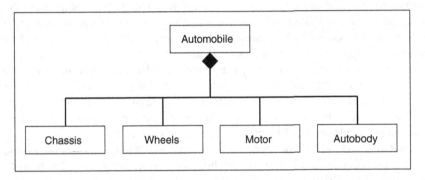

Fig. 6.9 Composition in UML

For a more detailed description of the object orientated techniques that are listed here and of the UML related representation, further reading is suggested, for example [Oest-06].

6.2 Additional Methods for Calculating of Flexibility

In Sect. 2.3, approaches to flexibility measurements of production systems have been analysed, according to chosen criteria that were considered to be especially appropriate. The aim was to gain thought-provoking impulses for a development that is consistent with the intention of the evaluation methodology, due to possible results. An additional overview of further evaluation methodologies of Volume-, Mix-, and Expansion flexibility was given. However, their weaknesses and strengths are not demonstrated explicitly, as they do not contribute to the insights of the evaluation methodologies that have been presented already. They are only outlined briefly. In the following they are ordered chronologically by the year of publication.

HOP suggests a methodology for measuring the ability of a production system to re-organize the production of one product variant to another in a quick and cost-efficient way. Here he refers to devices only, which he evaluates in regard to their number of jobs, their efficiency, their potential and their changeover time. He finally calculates an index of the device-related Mix flexibility with the help of the probability theory [Hop-04].

BRIEL evaluates the efficiency of adaptations in production systems with his scalable model. He estimates to what extent adaptation activities are advantagous, on the basis of estimating efforts, earnings, on the basis of capital formation and necessary investments. Although this is a monetary evaluation procedure only, conclusion about the Expansion flexibility of systems can be drawn. However, they assume stable assumptions of structure during the period of consideration [Brie-02].

KARSAK designed a procedure for capturing the Expansion flexibility of production systems by modifying the traditional present value method. Due to these modifications, specific abilities or functions of the expansion ability of a system become part of the evaluation of investment. On the basis of an expansion investment that has to be taken into consideration, with his procedure, KARSAK judges the consequences of investments that have to be expected as a result of not choosing the first investment alternative [Kars-02].

BENGTSSON and OLHAGER determine the Mix flexibility in production systems by evaluating existing, reasonable and useful production options under the consideration of capacity, setup activities, the degree of automation and the multiple usages of production resources. In this context they consider both theoretical and practical aspects of flexibility in a system [BeOl-02].

PARKER and WIRTH developed a framework which helps to carry out evaluations of the Volume- and Expansion flexibility of production systems. The basis for evaluating the Volume flexibility is a range of the output in which the considered system can produce economically and that has to be named in the run-up. However, in order to determine the Expansion flexibility, costs of a system expansion with a large dimension are compared with costs of several small expansions that are expected in case of missing great expansions [PaWi-99].

With his procedure, DAS allows to carry out measurements on Devices-, Work-plan-, Product- and Volume flexibility. Here he uses a special evaluation scale that is divided in the following criteria: demand, capability, effectiveness, inflexibility and optimality of systems. Coming from this, DAS identifies special characteristics at production systems that are allocated within the scale. After that they have to be evaluated in regard to the four kinds of flexibility that have to be determined [Das-96].

The procedure suggested by UPTON provides information on the ability of a production system to re-organize the production of one product variant to another. This form of Mix flexibility is called "mobility" and can be determined both for devices and for lines, at which response times of the changeover have to be considered. The less time needed for the setup, the higher the resulting index [Upto-95].

Classified by their timed mode of effectiveness in short-term and long-term flexibility, OST determines five kinds of flexibility due to specific evaluation criteria: Application-, Failure-, Integration-, Modification- and Volume flexibility. Using Fuzzy Logic he calculates definite indices for every kind of flexibility, by quantitatively relating imprecise interpretations and edge conditions. In addition to that, OST allows the use of specific user aspects of flexibility in form of cost cuttings or opportunity costs on the basis of static or dynamic procedure of investment costs [Ost-95].

FEKECS provides information of Mix flexibility on a workplace level of production systems with his quantitative evaluation procedure. For this he takes three decisive evaluation factors as a basis. They concern the potential application variety of executing different machining tasks, their speed to execute different set ups (under the inclusion of a setting time matrix), as well as costs that correspond to the tasks of set up [Feke-89].

ADAM, BACKHAUS, MEFFERT and WAGNER measure the Volume flexibility of a production system due to certain conditions/scenarios. In regard to the system they determine the best possible amount of output that is achievable with complete system flexibility only. Additionally they determine results of scenarios that are not optimal and which result from fixed output quantities and that are put in relation to the optimal output [ABMW-89].

The procedure of SON and PARK provides an analysis of the four different kinds of flexibility Product-, Process-, Expansion- and Volume flexibility. On the basis of cost-related criteria, they determine indices for each of them. Due to a reciprocal summation of single reciprocal indices, SON and PARK are able to name a standardized index for the Total flexibility of a system that can refer to both workplaces and lines [SoPa-87].

With his evaluation procedure, SCHÄFER uses the ratio of maximum achievable and pre-planned capacity of the system. The resulting determined degree of flexibility provides information about the Volume flexibility of a production system. Thus, he is able to judge to what extent the capacity of a production system increases or decreases in relation to standard capacity. SCHÄFER calculates the maximum available capacity that is needed for the evaluation from the minimum standard of capacity of the resources that are available in the point of view [Schä-80].

HANSMANN determines the flexibility of a production system under the assumption of detailed knowledge about future needs of development by setting the system in relation to a case of ideal adaptation. By involving total costs for realizing the resulting need of adaptation in the considered system, he determines the measured quantity F_j, that allows him to draw conclusions on the system specific Expansion flexibility [Hans-78].

6.3 Demonstration of Further Aspects Regarding the Evaluation Methodology

The cost-calculation frame of reference and the resulting need of information that have been defined in Sect. 3.3 are very complex. That is why in the following, further knowledge concerning the acquisition of workplace-related energy costs, as well as knowledge concerning the differentiation between process- and cycle time will be imparted.

6.3.1 Calculation of Energy Costs for Workplaces

The knowledge about product-related energy costs that occur at a workplace can be a cost-relevant fabrication criterion if the processing operation is energy intensive, such as laser welding. As in contrast to the other material costs of a product-workplace-combination (see Formula 3.34, p. 93), the determination of energy costs is more complicated and at this point their calculation is addressed. The kind of energy that is used for the fabrication, electric energy (electricity), chemical energy (fuel) or a combination of these two is irrelevant.

In order to determine the product-related energy costs for a workplace as correctly as possible, the process time $t_{PT}(MI,WP)$ is sub-divided in machining time (energy intensive), setup costs and required idle time costs (less energy intensive) according to the operation at the workplace (see Sect. 6.3.2). For this purpose, energy costs per time unit that are caused by the operation are multiplied with the processing time and summed up afterwards, as demonstrated by Formula 6.1. The resulting value has to be multiplied with the energy quotient q_{Energy} that includes cost differentials that are dependent on the time of day, as they could occur when electrical energy is used. Calculating the energy quotient is carried out by applying Formula 6.2.

Formula 6.1 Calculating of energy costs for a product-workplace-combination

$$K_{Energy}(MI, WP) = (t_{MT} \times k_{E-MT} + t_{SIT} \times k_{E-SIT}) \times q_{Energy},$$

where:

$K_{Energy}(MI,WP)$: Energy costs of a product-workplace-combination;

t_{MT}: Machining time of a product-workplace-combination;

t_{SIT}: Setup- and required idle time of a product-workplace-combination;

k_{E-MT}: Energy costs per time unit for machining operations of a product-workplace-combination;

k_{E-SIT}: Energy costs per time unit for the setup and required idle time of a product-workplace-combination;

q_{Energy}: Energy quotient to consider the time of day and the corresponding energy tariff of a product-workplace-combination

Formula 6.2 Calculating of the energy quotient

$$q_{Energy}(MI, WP) = \frac{K_{RR} \times T_{RR} + K_{SR} \times T_{SR}}{(T_{RR} + T_{SR}) \times K_{RR}},$$

whereas:

$q_{Energy}(MI,WP)$: Energy quotient of a product-workplace-combination;

K_{RR}: Energy costs per time unit at regular rate;

K_{SR}: Energy costs per time unit at saver rate;

T_{RR}: Uptime of a workplace *WP* that is needed to produce a product *MI* at regular rate time;

T_{SR}: Uptime of a workplace *WP* that is needed to produce a product *MI* at saver rate time

6.3.2 Process Time as Part of the Cycle Time

The *cycle time* of a product is that span of time that is required from the beginning of the machining until the completion. It consists of setup-, machining- and required idle time [REFA-78]. In detail these are:

- The *machining time* is the time segment that is part of the cycle time in which the manufacture of a product is actually carried out [BeWe-05].
- The *setup time* characterizes the time that is needed for all activities that involve the setup of production equipment for a certain process, as for example to setup a machine with the required cutting tools. This also concerns those activities that are necessary to reposition the production equipment to their original setup status. There is no productive use of a production facility during the setup as no product is generated [REFA-78] [REFA-84].

Time belonging to the required idle time is when the production procedure is interrupted or the product cannot be worked on, or the product cannot be transported, stored or checked. It is differentiated between required idle times und non required idle times/additional idle times. *Required idle times* result from specific production conditions that are unavoidable. Examples are producing with buffers, time that is needed to let the machine recover from heating or the like, or waiting- and transport time. *Additional idle times* refer to idle times that trace back to failures. Unscheduled breakdowns of machines, manual interruption of the production, energy loss etc. are some examples [REFA-78] [Nebl-07].

In connection with the development of the flexibility evaluation methodology, only the cycle time was considered, as additional idle time is considered separately (see Formula 3.5, p. 61). With this point of view the term *process time* is introduced. In contrast to cycle time it does not consider additional idle time, as demonstrated by Formula 6.3.

Formula 6.3 Calculating of process time

$$t_{PT}(MI, WP) = t_M + t_{Set} + t_{rIT},$$

where:

t_{PT} (*MI,WP*): Process time for a product-workplace-combination;

t_M: Machining time for a product-workplace-combination;

t_{Set}: Setup time for a product-workplace-combination;

t_{rIT}: Required idle time for a product-workplace-combination

6.4   Example of a Production System
(for Illustrating Flexibility Calculations)

An important basis for the conception of the evaluation methodologies that are presented in this book were two factors: considerable dialogues with experts of the production field and detailed analyzes of the considered field for defining requirements. This resulted in different special cases at production systems that had to be considered in order to achieve a successful concept realization. This is why a fictitious production system was designed that is orientated to these special cases. It was part of numerous tests during the conception phase and was used as an example in Sect. 3.2. In the following, this system will be described by the main characteristics that are necessary for calculating the flexibility that also include expansion activities.

6.4.1   Graphical Structure and Terms

The fictitious production system consists of the two segments *S1* and *S2*, at which segment *S1* contains a line *L1.1*, which again includes two workplaces *WP1.1.1* and *WP1.1.2*. In addition to these two workplaces, there is another workplace *WP1.0.1* within *S1*. However, it is not part of any line, but its higher-level system is *S1* too. In contrast, segment *S2* contains a line *L2.1* to which three workplaces *WP2.2.1*, *WP2.2.2* and *WP2.2.2* are related. The relations of the mentioned sub-systems are represented in Fig. 6.10, that illustrates the described connections with the help of a tree structure. The root of the tree is represented by the system object "factory". From this factory different subsystems spread according to the organizational hierarchy that was introduced in Sect. 2.1.3.

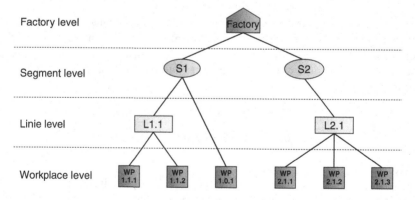

Fig. 6.10 Illustration of the fictitious production system in the form of a tree structure

The numeration of the subsystem is carried out according to the schema that is represented by Fig. 6.10. Segments get the term $S[x]$, $[x]$ is for the running number of the segment. Lines are described by $L[x].[y]$ and workplaces by $WP[x].[y].[z]$. Here $[y]$ stands for the number of the line in segment $S[x]$ and $[z]$ stands for the number of the workplace. The naming of subsystems is carried out on the basis of a running numeration, starting with 1. Thus, the first line in segment $S2$ gets the term $L2.1$, which leads to a term of $WP1.1.1$ for the first workplace in this line. However, workplace $WP1.0.1$ takes a $[y]$-value of 0 as it is not part of any line.

6.4.2 Calculating Parameters of the System

The fictitious production system produces six different kinds of products ($M = \{MI_1, \ldots, MI_6\}$). Four of them are part of the quantity of the final products ($FP = \{MI_2, MI_3, MI_4, MI_6\}$). Here it has to be considered that two of these products (MI_2, MI_4) are sellable intermediate products, because they are both components of another product and generate proceeds. The following Table 6.6 indicates the associated market value per quantity unit for each of the six manufactured goods. Furthermore, it contains a list for all products that shows all other products that are needed for its production. However, an additional registration of purchased materials does not occur, as they are already part of variable production costs.

The fabrication of the products takes place at one or more workplaces that are mentioned in Fig. 6.10. Possible product-workplace-combinations result herefrom (r_k) for which specific parameters of the production apply as shown in Table 6.7. They refer to a chosen working hour model WHM. The example deals with a regular shift model that consists of five workdays with 8 h each.

Table 6.6 Sales price and component conditions in the fictitious production system

Manufacturing item	MI_1	MI_2	MI_3	MI_4	MI_5	MI_6
S_{MIi} in MU/QU	–	4.00	11.00	3.00	–	10.00
Component	–	$1 \times MI_1$	$1 \times MI_1$	–	$2 \times MI_4$	$3 \times MI_5$
			$2 \times MI_2$			

Table 6.7 Cost- and non-cost-related calculation parameters for each product-workplace-combination in the fictitious production system

System object	$WP1.1.1$		$WP1.1.2$		$WP1.0.1$	$WP2.1.1$	$WP2.1.2$	$WP2.1.3$
M_{WP}	MI_1	MI_2	MI_2	MI_3	MI_4	MI_4	MI_5	MI_6
$C_{var}(MI, WP)$ in MU/QU	0.50	0.80	1.00	1.50	2.00	1.30	0.90	1.70
$a(MI, WP)$	1%	2%	1%	5%	1%	3%	2%	2%
$t_{PT}(EZG, WP)$	1 s	2 s	4 s	5 s	3 s	2 s	4 s	12 s
$t_{aIT}(WP)$	4,500 s		3,000 s		8,000 s	4,000 s	4,500 s	4,000 s
t_{max}	144,000 s							

Table 6.8 System object-related fixed costs (cost-related calculation parameters) for the fictitious production system

System object	WP1.1.1	WP1.1.2	WP1.0.1	WP2.1.1	WP2.1.2	WP2.1.3	Sum
$C_{Fix}(WP)$ in MU	250	310	380	450	200	460	**2,050**
System object	L1.1		–	L2.1			
$C_{Fix}(Linie)$ in MU	1,500			2,200			**3,700**
System object	S1			S2			
$C_{Fix}(Segment)$ in MU	6,900			5,800			**12,700**
System object	Fabrik						
$C_{Fix}(Factory)$ in MU	28,050						**28,050**

Besides the calculation parameters for the production system that are mentioned in Table 6.7, system object-related fixed costs $K_{Fix}(S)$ have to be gathered too, in order to complement the non-cost-related parameters. In Table 6.8 they are mentioned for each subsystem, and normalized to a special analysis period according to Formula 3.39 (p. 104). Different fixed costs are valid for the considered working time model, the regular shift model, only.

The sale ratio that is assumed for the final products in the fictitious production system is:

$$MI_2 : MI_3 : MI_4 : MI_6 = 1 : 2 : 1.5 : 2$$

Thus, the resulting product mix vector v is:

$$v = (0 \times MI_1, 1 \times MI_2, 2 \times MI_3, 1.5 \times MI_4, 0 \times MI_5, 2 \times MI_6)^T$$
$$= (0; 1; 2; 1.5; 0; 2)^T$$

6.4.3 Expansion Alternatives for Segment S2

In Sect. 3.2.4 different examples for demonstrating the procedure of calculating the Expansion flexibility were introduced. In the following three alternatives for capacity expansion are mentioned that relate to those examples. They concern segment S2 only.

6.4.3.1 Alternative 1: Construction of a Redundant Production Line

In this alternative a new line L2.2 that is redundant to line L2.1 is built up in segment S2 and is illustrated in the style of Fig. 6.10:

For reasons of convenience, line L2.2 that is represented in Fig. 6.11 features the same cost- and non-cost-related calculation parameters as line L2.1 (see Table 6.7, p. 169 and Table 6.8, p. 170). However, in this case additional expansion costs of $C_{Exp,P}(S_2) = 150$ MU per period have to be calculated for segment S2.

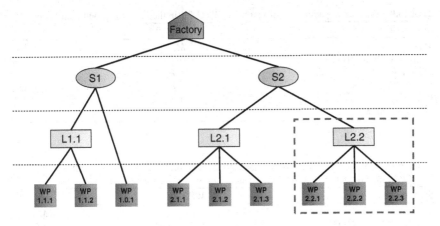

Fig. 6.11 Expansion of the fictitious production system by line L2.2

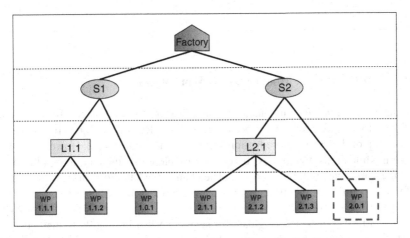

Fig. 6.12 Expansion of the fictitious production system by workplace WP2.0.1

6.4.3.2 Alternative 2: Construction of an Additional Workplace

In this scenario, segment *S2* is expanded by an additional workplace *WP2.0.1* that can produce both MI_5 and MI_6. The production of MI_4 is not essential as its additional production can be carried out by existing manufacture facilities in segment *S1* to a certain degree. Figure 6.12 shows the appropriate modifications to the example production system.

For this new workplace *WP2.0.1* the cost- and non-cost-related calculation parameters that are mentioned in Table 6.9 are valid, all other marginal conditions are the same.

Additionally, with the setup of the new workplace *WP2.0.1* expansion costs of $K_{Exp,P}(S_2) = 100$ MU per period arise for segment *S2*.

Table 6.9 Calculation parameters of workplace *WP2.0.1* in the fictitious production system

System object	WP2.0.1	
M_{WP}	MI_5	MI_6
$C_{var}(MI, WP)$ in MU/QU	0.90	1.70
$a(MI, WP)$	3%	3%
$t_{PT}(MI, WP)$	6 s	15 s
$t_{aIT}(WP)$	4,000 s	
$C_{Fix,P}(WP)$ in MU	600	

Table 6.10 Changed cost and non-cost-related calculation parameters for workplaces *WP2.1.2* and *WP2.1.3* in the fictitious production system

System object	WP2.1.2	WP2.1.3
M_{WP}	MI_5	MI_6
$C_{var}(MI, WP)$ in MU/QU	0.85	1.65
$a(MI, WP)$	1.5%	1.5%
$t_{PT}(MI, WP)$	3.5 s	10.5 s
$t_{aIT}(WP)$	4,000 s	3,500 s
$C_{Fix}(WP)$ in MU	230	490
t_{max}	144,000 s	

6.4.3.3 Alternative 3: Modification of Workplaces

Expansion alternative 3 includes the modification of workplaces *WP2.1.2* and *WP2.1.3* by adding additional processing modules. Regarding Fig. 6.10, the structural setup of the example production system remains unchanged. However, due to these modifications, the process time, the variable costs, the additional idle times and the scrap rates will be improved. The changed calculation parameters that are effective for the example production system are listed in Table 6.10.

Due to the expansion activities at the two workplaces, expansion costs of $K_{Erw,P}(S_2) = 50$ MU per period originate for segment S2. Additionally to that, costs of production equipment of the two workplaces increase due to the acquirement of additional process modules. Hence their part of fixed costs increases by 30 MU per period.

6.5 Calculation Parameters of the Expansion Activities from the Example Production System

The following cost and non-cost-related calculation parameters refer to different system objects that were part of the expansion activities that were introduced in Sect. 4.2.3.2. They were carried out to correct the flexibility deficits that were identified in the example production system (see Tables 6.11–6.13).

Table 6.11 Calculation parameters of the workplaces that are needed to correct flexibility deficits in the case study

System object	VT.L2.A_7 (new)	VT.L2.B_8 (new)	VT.L2.A_7 (new50%)	VT.L2.AB_9 (new)		EP.0.AB_3 (new)	
M_{WP}	VT-A1	VT-B1	VT-A1	VT-A1	VT-B1	A_3	B_3
$a(MI, WP)$	3%	3%	3%	3%	3%	1%	1%
$t_{PT}(MI, WP)$	294 s	294 s	588 s	330 s	330 s	180 s	150 s
$C_{var}(MI, WP)$	€1.26	€1.20	€1.26	€1.40	€1.35	€42.00	€38.63
$C_{Fix}(WP)$ per week	€438.92	€438.92	€325.00	€404.70		€518.17	
$t_{max}(WHM)$	$WHM_1 = 144,000$ s; $WHM_2 = 180,000$ s; $WHM_3 = 208,800$ s; $WHM_4 = 288,000$ s; $WHM_5 = 345,600$ s; $WHM_6 = 403,200$ s						
$t_{aWT}(WP)$	2% $t_{max}(WHM)$						

Table 6.12 Alternative dependent expansion costs of line "*VT.L2*" (normalized to one consideration period)

Expansion alternatives	Expansion costs of line "*VT.L2*" per period
Alternative 1 Construction of a redundant workplace *VT.L2.A_7(new)* and *VT.L2.B_8(new)*	€1,675
Alternative 2 Construction of a redundant workplace *VT.L2.B_8(new)* and a redundant workplace *VT.L2.A_7(new50%)* with 50% capability	€1,122
Alternative 3 Construction of a multiple-product manufacturing *workplace VT.L2.AB_9(new)*	€848

Table 6.13 Expansion costs of segment "Verbauteile" (normalized to one consideration period)

Expansion alternative	Expansion costs of segment "Verbauteile"
VT.L2.A_7(new) + *VT.L2.B_8(new)*	€750

References

[ABMC-05]   Alexopoulos K, Burkner S, Milionis I, Chryssolouris G (2005) DESYMA – an integrated method to aid the design and the evaluation of reconfigurable manufacturing systems. In: Proceedings of the 1st International Conference on Changeable, Agile, Reconfigurable and Virtual Production (CARV 2005), Munich, Germany

[ABMW-89]   Adam D, Backhaus K, Meffert H, Wagner H (1989) Integration und Flexibilität – Eine Herausforderung für die Allgemeine Betriebswirtschaftlichkeitslehre. Gabler Verlag, Wiesbaden

[Adam-98]   Adam D (1998) Produktions-management, vol 9, Auflage. Gabler Verlag, Wiesbaden

[Aggt-87]   Aggteleky B (1987) Fabrikplanung – Werksentwicklung und Betriebsrationalisierung, Band 1, vol 2, Auflage. Carl Hanser Verlag, München

[Aggt-90]   Aggteleky B (1990) Fabrikplanung – Werksentwicklung und Betriebsrationalisierung, Band 2, vol 2, Auflage. Carl Hanser Verlag, München

[AIKT-04]   Arnold D, Isermann H, Kuhn A, Tempelmeier H, Furmans K (Hrsg.) (2004) Handbuch Logistik. 3. Auflage. Springer Verlag, Berlin, Heidelberg u.a.

[AKN-06]   Abele E, Kluge J, Näher U (2006) Handbuch globale produktion. Hanser Verlag, München

[AlSe-06]   Ali SA, Seifoddini H (2006) Simulation intelligence and modeling for manufacturing uncertainties. In: Proceedings of the 2006 Winter Simulation Conference

[AlSS-05]   Ali SA, Seifoddini H, Sun H (2005) Intelligent modeling and simulation of flexible assembly systems. In: Proceedings of the 2005 Winter Simulation Conference

[AMMC-05]   Alexopoulos K, Mamassioulas A, Mourtzis D, Chryssolouris G (2005) Volume and product flexibility: a case study for a refrigerators producing facility. In: Proceedings of the 10th IEEE International Conference on Emerging Technologies and Factory Automation (ETFA 2005). Catania/Italy

[AMPC-07]   Alexopoulos K, Mourtzis D, Papakostas N, Chryssolouris G (2007) DESYMA – assessing flexibility for the lifecycle of manufacturing systems. Int J Prod Res 45 (7):1683–1694

[APMG-06]   Alexopoulos K, Papakostas N, Mourtzis D, Gogos P, Chryssolouris G (2006) Quantifying the flexibility of a manufacturing system by applying the transfer function. Int J Comput Integrated Manuf 20:538–547

[ARKO-07]   Abul Ola, H, Rogalski S, Krahtov K, Ovtcharova J (2007) Change management for the production of the future. 2. International Conference on Changeable, Agile, Reconfigurable and Virtual Production (CARV 2007), Toronto/Canada

[Balz-00]   Balzert H (2000) Objektorientierung in 7 Tagen – Vom UML-Modell zur fertigen web-Anwendung. Spektrum Akademischer Verlag GmbH, Berlin

S. Rogalski, *Flexibility Measurement in Production Systems*,
DOI 10.1007/978-3-642-18117-7, © Springer-Verlag Berlin Heidelberg 2011

[Bart-05]	Barth H (2005) Produktionssysteme im Fokus. In: wt Werkstattstechnik online, Jahrgang 95, Heft 4, Springer-VDI-Verlag, Düsseldorf
[BBBD-03]	Bloech J, Bogaschewsky R, Buscher U, Daub A, Götze U, Roland F (2003) Einführung in die produktion, vol 5, überarbeite Auflage. Physica Verlag, Heidelberg
[BCS-07]	Blank A, Christ H, Schneider K-H (2007) Betriebswirtschaftslehre, vol 3, Auflage, Fachhochschule für Wirtschaft. Bildungsverlag EINS, Troisdorf
[BeHö-05]	Benz J, Höflinger M (2005) Logistikprozesse mit SAP R/3: eine anwendungsbezogene Einführung – mit durchgehendem Fallbeispiel. Vieweg Verlag, Wiesbaden
[Behr-85]	Behrbohm P (1985) Flexibilität in der industriellen produktion: Grundüberlegungen zur Systematisierung und Gestaltung der produktionswirtschaftlichen Flexibilität. Verlag Lang, Frankfurt a. M
[Bell-05]	Bellmann K (2005) Flexibilisierung der Produktion durch Dienstleistungen. In: Kaluza B, Blecker T (Hrsg.) Erfolgsfaktor Flexibilität. Erich Schmidt Verlag, Berlin
[BeLu-03]	Becker J, Luczak H (2003) Workflowmanagement in der Produktionsplanung und – steuerung. Springer Verlag, Berlin
[Benj-94]	Benjaafar S (1994) Models for performance evaluation of flexibility in manufacturing systems. Int J Prod Res 32(6):1383–1402
[BeOl-02]	Bengtsson J, Olhager J (2002) Valuation of product-mix flexibility using real options. Int J Prod Econ 78:13–28
[BeWe-05]	Best E, Weth M (2005) Geschäftsprozesse optimieren: der Praxisleitfaden für erfolgreiche Reorganisation, vol 2, überarb. Auflage. Gabler Verlag, Wiesbaden
[Beye-04]	Beyer H-T (2004) Optimales Lieferantenmanagement. Online Lehrbuch Marktprozesse, Universität Erlangen-Nürnberg, http://www.economics.phil.uni-erlangen.de/bwl/lehrbuch/kap2/liefmgt/ liefmgt.pdf, Stand: 16.08.2007
[BiSc-95]	Bichler K, Schröter N (1995) Praxisorientierte logistik. Kohlhammer Verlag, Stuttgart
[Blei-04]	Bleicher K (2004) Das Konzept integriertes management. Visionen – Missionen – Programme, vol 4, Auflage. Campus Verlag, Frankfurt/M
[BMF-01]	Bundesministerium der Finanzen (2001) Abschreibungstabellen. Berlin
[BMJ-09]	Bundesministerium der Justiz: § 4 Straßenbahnen, Obusse, Kraftfahrzeuge. http://bundesrecht.juris.de/pbefg/__4.html, Stand: 27.02.2009
[BrGr-04]	Braßler A, Grau C (2004) Modulare organisationseinheiten. Arbeits- und Diskussionspapiere der Wirtschaftswissenschaftlichen Fakultät der Universität Jena, 25/2004
[Brie-02]	von Briel R (2002) Ein skalierbares Modell zur Bewertung der Wirtschaftlichkeit von Anpassungsinvestitionen in ergebnisverantwortlichen Fertigungssystemen. Dissertation, Universität Stuttgart
[Brum-94]	Brumberg C (1994) Zeitliche flexibilisierung im industriebetrieb: analyse und ansätze zum Abbau organisatorsicher und verhaltensbedingter restriktionen. Gabler Verlag, Wiesbaden
[Bühn-04]	Bühner R (2004) Personalmanagement, vol 3, Auflage. Oldenbourg Verlag, München
[Bunz-85]	Bunz A (1985) Strategieunterstützungsmodelle für Montageplanungen. In: SzU, Band 11. Verlag Lang, Frankfurt a. M.
[Burd-02]	Burdelski T (2002) Rechnungswesen I: Übungen, Lösungen zur Vorlesung Rechnungswesen. Skript zur Vorlesung, Universität Karlsruhe
[Chry-05]	Chryssolouris G (2005) Manufacturing systems-theory and practice, 2nd edn. Springer-Verlag, Berlin
[Chry-96]	Chryssolouris G (1996) Flexibility and its measurement. CIRP Annals – Manufacturing Technology, 45(2):581–587

References

177

[Coen-03]	Coenenberg AG (2003) Kostenrechnung und Kostenanalyse, vol 5, Auflage. Schäffer-Poeschel Verlag, Stuttgart
[Cors-99]	Corsten H (1999) Produktionswirtschaft, vol 8, Auflage. Oldenbourg Wissenschaftsverlag, München u.a
[CQP-89]	Corlett EN, Queinnec Y, Paoli P (1989) Die Gestaltung der Schichtarbeit. In: Europäische Stiftung zur Verbesserung der Lebens- und Arbeitsbedingungen (Hrsg.) Informationsbroschürenserie Nr. 8. Amt für Amtliche Veröffentlichungen der Europäischen Gemeinschaften, Luxemburg
[CSS-99]	Calic M, Sieling B, Simon P (1999) Grundbegriffe der Objektorientierung. In: Objektorientierung – State-of-the-Art. Hüthig Verlag, Heidelberg, pp S.7–S.22
[Dang-03]	Dangelmaier W (2003) Produktion und information. System und modell. Springer Verlag, Berlin, Heidelberg
[Das-96]	Das SK (1996) The measurement of flexibility in manufacturing systems. Int J Flex Manuf Syst 8:67–93
[DCR-07]	Dyckhoff H, Clermont M, Rassenhövel S (2007) Industrielle Dienstleistungsproduktion. In: Corsten H, Missbauer H (Hrsg.) Produktions- und Logistikmanagement. Oldenbourg Wissenschaftsverlag, München
[DeTo-98]	De Toni A, Tonchia S (1998) Manufacturing flexibility: a literature review. Int J Prod Res 36(6):1587–1617
[DHJM-06]	Denkena B, Harms A, Jacobsen J, Möhring H-C, Jungk A, Noske H (2006) Lebenszyklus-orientierte Werkzeugmaschinenentwicklung. In: wt Werkstatttechnik online, Jahrgang 96, Heft 7/8, S. 441–446, Springer-VDI-Verlag, Düsseldorf
[DIN310-51]	DIN 31051 (Deutsche Industrie Norm) (2003) Grundlagen der Instandhaltung. Ausgabe 2003–06, Beuth Verlag
[DIN67-89]	DIN 6789 (Deutsche Industrie Norm): Dokumentationssystematik. Aufbau technischer Produktdokumentationen. Beuth Verlag, Berlin 1993.
[DoDr-07]	Domschke W, Drexl A (2007) Einführung in operations research, vol 7, Auflage. Springer Verlag, Berlin
[DoQu-04]	Dombrowski U, Quack S (2004) Die ungenutzten Potentiale in bestehenden Fabriken. In: 5. Deutsche Fachkonferenz Fabrikplanung, Tagungsband, 31 März–1 April 2004, Stuttgart
[Dorm-86]	Dormmayer H-J (1986) Konjunkturelle Früherkennung und Flexibilität im Produktionsbereich. In: Beiträge zur quantitativen Wirtschaftsforschung, Bd. 3, München
[Dumk-00]	Dumke R (2000) Software engineering, vol 2, Auflage. Vieweg Verlag, Braunschweig, Wiesbaden
[DuOe-93]	Dunckel H, Oesterreich R (1993) Mensch – Technik – Organisation 5a. Kontrastive Aufgabenanalyse im Büro. Der KABA Leitfaden. Grundlagen und Manual, Verlag der Fachvereine, Zürich
[Dürr-00]	Dürrschmidt S (2000) Planung und Betrieb wandlungsfähiger Logistiksysteme in der variantenreichen Serienproduktion. Dissertation TU München
[Eber-04]	Ebert G (2004) Kosten- und Leistungsrechnung, vol 10, Auflage. Gabler Verlag, Wiesbaden
[EBGK-02]	Eversheim W, Bergholz M, Gather M, Kerner S, Lange-Stalinski T, Lanza M, Laufenberg L, Lay K, Reinfelder A, Schreiber W, Sulser HP (2002) Die Fabrik von morgen: vernetzt und wandlungsfähig!. In: AWK Aachener Perspektiven: Wettbewerbsfaktor Produktionstechnik, Shaker Verlag Aachen, S.73–S.96
[Ever-89]	Eversheim W (1989) Organisation in der Produktionstechnik. Band 4: Fertigung und Montage, 2. Auflage. VDI Verlag, Düsseldorf
[Ever-92]	Eversheim W (1992) Flexible Produktionssysteme. In: Frese E (Hrsg.) Handwörterbuch der Organisation. Poeschel-Verlag, Stuttgart

[EvSc-99]   Eversheim W, Schuh G (1999) Betrieb von Produktionssystemen, vol 4, Produk-
              tion und Management. Springer Verlag, Berlin u.a
[Feke-89]   Fekecs B (1989) Ein Ansatz zur quantitativen Bewertung der Flexibilität von
              Fertigungssystemen. In: wt Werkstattstechnik 79, S.601–S.604
[Feld-91]   Feldhahn K-A (1991) Logistikmanagement in kleinen und mittleren Unternehmen.
              Dissertation TU Braunschweig
[FHPP-99]   Fandel G, Heuft B, Paff A, Pitz T (1999) Kostenrechnung. Springer Verlag,
              Berlin, Heidelberg, New York
[Frau-05]   Frauenfelder P (2005) Produktions-Management in der Unternehmensführung.
              UF Wintersemester 2005/06 V-06(14), Universität Zürich (ETH), Zürich
[Geye-06]   Geyer H (2006) Praxiswissen BWL. Haufe Verlag DE, Freiburg
[Götz-04]   Götze U (2004) Kostenrechnung Und Kostenmanagement. Springer Verlag,
              Berlin
[GPMC-07]   Georgoulias K, Papakostas N, Makris S, Chryssolouris G (2007) A toolbox
              approach for flexibility measurements in diverse environments. CIRP Annals
              Manuf Technol 56(1):423–426
[GüHa-99]   Günthner WA, Haller M (1999) Im Spannungsfeld zwischen Flexibilität und
              Automatisierung. In: Hossner R (Hrsg.) Jahrbuch Logistik 1999. Verlagsgruppe
              Handelsblatt GmbH, Düsseldorf
[Günt-05]   Günthner WA (2005) Anpassungssituationen im automobilen Netzwerk – Eine
              Wertung der Akteure. In: Industrie Management 21 (2005) 5, GITO-Verlag,
              Berlin
[Gute-83]   Gutenberg E (1983) Grundlagen der Betriebswirtschaftslehre – Band 1: Die
              Produktion. Springer Verlag, Berlin
[Hall-99]   Haller M (1999) Bewertung der Flexibilität automatisierter Materialflusssysteme
              der variantenreichen Großserienproduktion. Dissertation TU München
[Hans-06]   Hansmann KW (2006) Industrielles management, vol 8, überarbeitete Auflage.
              Oldenbourg Wissenschaftsverlag, München
[Hans-78]   Hanssmann F (1978) Einführung in die Systemtechnik. Oldenbourg Verlag,
              München
[HaWh-88]   Hayes RH, Wheelwright SC, Clark KB (1988) Dynamic manufacturing – creating
              the learning. Organization, New York, London
[Hell-08]   Hellert U (2008) Praxis der Nacht- und Schichtplangestaltung. LIT Verlag,
              Berlin, Hamburg, Münster
[HiUl-79]   Hill W, Ulrich P (1979) Wissenschaftliche Aspekte ausgewählter betriebswis-
              senschaftlicher Konzeptionen. In: Raffee H, Abel B (Hrsg.) Wissenschaftstheo-
              retische Grundlagen der Wirtschaftswissenschaften. Vahlen Verlag, München
[Hofm-04]   Hofman I (2004) Kostenrechnung "light". Wirtschaft, Recht, Mitbestimmung,
              Fernlehrgang des Österreichischen Gewerkschaftsbundes, Wien
[HoMa-86]   Horvárth P, Mayer R (1986) Produktionswirtschaftliche Flexibilität. In:
              Wirtschaftswissenschaftliches studium, vol 2, Heft., pp S. 69–S. 76
[Hop-04]   Hop NV (2004) Approach to measure the mix response flexibility of
              manufacturing systems. Int J Prod Res 42(7):1407–1418
[Hopf-89]   Hopfmann L (1989) Flexibilität im Produktionsbereich – Ein dynamisches mod-
              ell zur analyse und Bewertung von Flexibilitätspotentialen. Verlag Peter Lang,
              Frankfurt a. M
[Jaco-74]   Jacob H (1974) Unsicherheit und Flexibilität: Zur Theorie der Planung bei
              Unsicherheit. In: Zeitschrift für Betriebswirtschaft, 44. Jg., S. 229–326,
              403–448 und 505–526
[KaBl-05a]   Kaluza B, Blecker T(2005) Flexibilität – State of the Art und Entwicklungstrends.
              In: Kaluza B, Blecker T (Hrsg.) Erfolgsfaktor Flexibilität. Erich Schmidt Verlag,
              Berlin

[KaBl-05b]   Kaluza B, Blecker T (2005) Erfolgsfaktor Flexibilität. Strategien und Konzepte für wandlungsfähige Unternehmen. Erich Schmidt Verlag, Berlin

[Kale-04]   Kalenberg F (2004) Grundlagen der Kostenrechnung. Eine anwendungsorientierte Einführung. Oldenbourg Verlag, München

[Kalu-93]   Kaluza B(1993) Flexibilität, betriebliche. In: Grochla E, Wittmann W (Hrsg.) Handwörterbuch der Betriebswirtschaft. Schäffer- Poeschel Verlag, Stuttgart

[Kars-02]   Karsak EE (2002) An options approach to valuing expansion flexibility in flexible manufacturing system investments. Eng Econ 47(2):S. 169–S. 189

[KBA-09]   Kraftfahr-Bundesamt: Fahrzeugklassen und Aufbauarten. http://www.kba.de/cln_005/nn_191224/DE/Statistik/Fahrzeuge/Besitzumschreibungen/Fahrzeugk lassenAufbauarten/2008__u__fzkl__eckdaten__absolut.html, Stand: 27.02.2009

[KeBo-98]   Kemmetmüller W, Bogensberger S (1998) Handbuch der Kostenrechnung, vol 5, aktualisierte und erweiterte Auflage. Service Fachverlag, Wien

[Kees-98]   Kees A (1998) Objektorientierte PPS-Systementwicklung. In: Luczak H, Eversheim W (Hrsg.), Schotten M. Produktionsplanung und -steuerung. Springer-Verlag, Berlin u.a., S.653–S.695

[KeKe-05]   Kersten W, Kern E-M (2005) Flexibilität in der verteilten Produktentwicklung. In: Kaluza B, Blecker T (Hrsg.) Erfolgsfaktor Flexibilität. Erich Schmidt Verlag, Berlin

[Kern-80]   Kern W (1980) Industrielle Produktionswirtschaft, vol 3, überarbeitete Auflage. Poeschel Verlag, Stuttgart

[Kich-98]   Kicherer H (1998) Kosten- und Leistungsrechnung. Verlag C. H. Beck, München

[Knof-91]   Knof H-L (1991) CIM und organisatorische Flexibilität. Dissertation Universität Bochum

[Kobe-08]   Kober M (2008) Fertigungssegmentierung. MKP Produktionsgestaltung http://www.mkpro.de/fertigung.html, Stand: 21.04.2008

[Kohl-07]   Kohler U (2007) Methodik zur kontinuierlichen und kostenorientierten Planung produktionstechnischer systeme. Herbert Utz Verlag, München

[KoKr-08]   Korves B, Krebs P (2008) Bewertung und Planung von Fabriken unter Flexibilitätsgesichtspunkten bei der Siemens AG. In: Münchener Kolloquium-Innovationen für die produktion. Herbert Utz Verlag, München, pp S.57–S.68

[KoMa-99]   Koste LJ, Malhorta MK (1999) A theoretical framework for analysing the dimensions of manufacturing flexibility. J Oper Manage 18:S.75–S.93

[Krop-01]   Kropp W (2001) Systemische Personalwirtschaft, vol 2, Auflage. Oldenbourg Verlag, München

[KRS-06]   Krappe H, Rogalski S, Sander M (2006) Challenges for handling flexibility in the change management process of manufacturing systems. IEEE Conference on Automation Science and Engineering (IEEE-CASE), Shanghai

[KSW-02]   Klauke A, Schreiber W, Weißner R (2002) Zukunftsorientierte Fabrikstrukturen in der Automobilindustrie. In: wt Werkstattstechnik online, Jahrgang 92, Heft 4, Springer-VDI-Verlag, Düsseldorf

[Kühn-89]   Kühn M (1989) Flexibilität in logistischen systemen. Physica, Heidelberg

[Lenz-04]   Lenz B (2004) Verkettete Orte: Filieres in der Blumen- und Zierpflanzenproduktion, vol 1, Auflage. Lit-Verlag, Münster u.a

[LES-06]   Lechner K, Egger A, Schauer R (2006) Einführung in die Allgemeine Betriebswirtschaftslehre, vol 23, Auflage. Linde Verlag, Wien

[Lind-05]   Lindemann U (2005) Der Änderungsmanagement Report. CiDaD Working Paper Series, Jahrgang 1, Nr. 1 ISSN 1861–079X, August 2005

[Lucz-93]   Luczak H (1993) Arbeitswissenschaft. Springer Verlag, Berlin u. a

[Lutz-96]   Lutz B u.a. (Hrsg.) (1996) Produzieren im 21. Jahrhundert: Herausforderungen für die deutsche Industrie. Ergebnisse des Expertenkreises Zukunftsstrategien. Band 1, Frankfurt am Main

[LWW-00]   Lutz S, Windt K, Wiendahl H-P (Hrsg.) (2000) Produktionsmanagement in
             Unternehmensnetzwerken; Reihe:Wandelbare Produktionsnetze. Band 2, Verlag
             Praxiswissen, Dortmund
[MeBo-05]   Mertens P, Bodendorf F (2005) Programmierte Einführung in die Betriebs-
             wirtschaftslehre: Institutionenlehre, vol 12, überarb Auflage. Gabler Verlag,
             Wiesbaden
[Meff-68]   Meffert H (1968) Die Flexibilität in betriebswirtschaftlichen Entscheidungen.
             Unveröffentlichte Habilitationsschrift, München
[MEK-05]   Marr R, Elbe M, Kaduk S (2005) Arbeitszeitflexibilisierung - Grundlegendes
             Problem oder Erfolgsmodell moderner Arbeitsbeziehungen? In: Kaluza B,
             Blecker T (Hrsg.) Erfolgsfaktor Flexibilität. Erich Schmidt Verlag, Berlin
[MHE-03]   Meffle G, Heyd R, Weber P (2003) Das Rechnungswesen der Unternehmung als
             Entscheidungsinstrument. Band 1, 3. überarb. Auflage, Fortis Verlag
[Milb-02]   Milberg J (2002) Erfolg in Netzwerken. In: Milberg J, Schuh G (Hrsg.) Erfolg in
             Netzwerken. Springer Verlag, Berlin u.a., S. 5–16
[Nebl-07]   Nebl T (2007) Produktionswirtschaft, vol 6, überarb. und erw. Auflage. Oldenbourg
             Wissenschaftsverlag, München, Wien
[NeMo-04]   Neumann K, Morlock M (2004) Operations research, vol 2, Auflage. Hanser
             Verlag, München
[Neuh-01]   Neuhausen J (2001) Methodik zur Gestaltung modularer Produktionssysteme für
             Unternehmen der Serienproduktion. Dissertation RWTH Aachen
[Neum-96]   Neumann K (1996) Produktions- und operations-management. Springer Verlag,
             Berlin
[Niem-07]   Niemann J, (2007) Eine Methodik zum dynamischen Life Cycle Controlling von
             Produktionssystemen. Dissertation Universität Stuttgart
[NN-05]   N.N.: Kostenrechnung. Universität Klagenfurt, Sommersemester 2005, http://www.
             uni-klu.ac.at/csu/downloads/KORE_V_SS2005_Teil2.pdf, Stand: 19.12.2007
[Oest-06]   Oestereich B (2006) Analyse und design mit UML 2.1 – objektorientierte soft-
             wareentwicklung, vol 8, aktualisierte Auflage. Oldenbourg Wissenschaftsverlag,
             München
[Olfe-05]   Olfert K (2005) Kostenrechnung, vol 14, Auflage. Kiehl Friedrich Verlag,
             Ludwigshafen, Rhein
[ORK-07]   Ovtcharova J, Rogalski, S, Krahtov K (2007) eHomeostasis. Methodology in the
             Automotive Industry. MCPC 2007, World Conference on Mass Customization &
             Personalization (MCP), MIT Cambridge/Boston
[Ost-95]   Ost S (1995) Wirtschaftliche Bewertung der Produktionsflexibilität. Kostenrech-
             nungspraxis Z Controlling 3:S.153–S.158
[PaWi-99]   Parker R, Wirth A (1999) Manufacturing flexibility: measures and relationships.
             Eur J Oper Res 118(3):429–449
[PBS-05]   Peters S, Brühl R, Stelling JN (2005) Betriebswirtschaftslehre, vol 11, Auflage.
             Oldenbourg Verlag, München, Wien
[Penr-95]   Penrose E (1995) The theory of the growth of the firm, vol 3, Auflage. Oxford
             University Press, New York
[PeRu-01]   Peláez-Ibarrondo J, Ruiz-Mercader J (2001) Measuring operational flexibility.
             Manufacturing information systems proceedings of the fourth SMESME interna-
             tional conference, Stimulating Manufacturing Excellence in Small & Medium
             Enterprises, Aalborg/Denmark
[Petr-06]   Petry T (2006) Netzwerkstrategie: Kern eines integrierten managements von
             Unternehmungsnetzwerken. Gabler Verlag, Wiesbaden
[Phil-02]   Philippson C (2002) Koordination einer standortbezogen verteilten Produktions-
             planung und – steuerung auf der Basis von Standard-PPS-Systemen. Dissertation
             RWTH Aachen

[Pibe-01]   Pibernik R (2001) Flexibilitätsplanung in Wertschöpfungsnetzwerken. in: ZfB 71.Jg., Heft 8

[PlRe-06]   Plinke W, Rese M (2006) Industrielle Kostenrechnung, vol 7, Auflage. Springer Verlag, Berlin, Heidelberg

[REFA-78]   REFA (1978) Verband für Arbeitsstudien u. Betriebsorganisation: Methodenlehre des Arbeitsstudiums. 2. Teil Datenermittlung. 6. Auflage. Carl Hanser Verlag, München

[REFA-84]   REFA (1984) Verband für Arbeitsstudien u. Betriebsorganisation: Methodenlehre des Arbeitsstudiums. 1. Teil Grundlagen. 7. Auflage. Carl Hanser Verlag, München

[REFA-90]   REFA (1990) Verband für Arbeitsstudien und Betriebsorganisation (Hrsg.): Planung und Betrieb komplexer Produktionssysteme. Hanser Verlag, München

[Rein-02]   Reinhardt G (2002) Wandlungsfähige Fabrikgestaltung. in: ZWF Zeitschrift für wissenschaftlichen Fabrikbetrieb, Band 97, Heft 1–2

[Rein-04]   Reinhart G (2004) Flexibilität und Wandlungsfähigkeit von Fabriken im globalen Wettbewerb. In: 5. Deutsche Fachkonferenz Fabrikplanung, Tagungsband, 31 März–1 April 2004, Stuttgart

[RHQJ-05]   Rupp C, Hahn J, Queins S, Jeckle M, Zengler B (2005) UML 2 glasklar: Praxiswissen für die UML-Modellierung und -Zertifizierung. Hanser Verlag, München, Wien

[Rieb-94]   Riebel P (1994) Einzelkosten- und Deckungsbeitragsrechnung. Grundfragen einer markt- und entscheidungsorientierten Unternehmensrechnung, vol 7, Auflage. Gabler Verlag, Wiesbaden

[Roga-10]   Rogalski S (2010) Flexibilitätsbewertungswerkzeug ecoFLEX beim mittelständischen Produktionsbetrieb im Einsatz, wt Werkstattstechnik online, Springer-VDI-Verlag, Jahrgang 100 (2010) H. 3, S.120–S.124

[Röhr-03]   Röhrs A (2003) Produktionsmanagement in Produktionsnetzwerken. Dissertation Universität Siegen; Europäischer Verlag der Wissenschaften, Frankfurt a. M.

[RoKr-06a]   Rogalski S, Krahtov K (2006) Änderungsmanagement für zukunftsorientierte Produktion (Teil 1). eDM-Report Nr. 2, Dressler Verlag

[RoKr-06b]   Rogalski S, Krahtov K (2006) Änderungsmanagement für zukunftsorientierte Produktion (Teil 2). eDM-Report Nr. 3, Dressler Verlag

[RoOv-09a]   Rogalski S, Ovtcharova J (2009) ecoFLEX – inter-branch methodology for flexibility measurements of production systems, presentation and proceeding to flexible automation and intelligent manufacturing, 19th international conference (FAIM 2009), 6–8 July 2009, Middlesbrough, UK

[RoOv-09b]   Rogalski S, Ovtcharova J (2009) Flexibilitätsbewertung von Produktionssystemen – ecoFLEX- eine branchenübergreifende Methodik, Zeitschrift für wissenschaftlichen Fabrikbetrieb (ZWF), 104. Jahrgang, Hanser Verlag, Heft 1–2/ 2009, S.64–S.70, München

[Ropo-99]   Ropohl G (1999) Allgemeine Technologie- Eine Systemtheorie der Technik, vol 2, Auflage. Carl Hanser Verlag, München, Wien

[RSO-09]   Rogalski S, Siebel J, Ovtcharova J (2009) EcoFLEX – introduction of a novel methodology for inter-branch flexibility measurement of production systems, proceeding to 2nd International Multi-Conference on Engineering and Technological Innovation: IMETI 2009, July 10th–13th, 2009 Orlando, Florida

[RüSt-00]   Rüttgers M, Stich V (2000) Industrielle Logik. Wissenschaftsverlag Mainz in Aachen, Achen

[ScBe-03]   Schuh G, Bergholz M (2003) Collaborative production on the basis of object oriented. Software Engineering Principles, CIRP Annals, Band 52, Heft 1

[Schä-80]   Schäfer F-W (1980) System zur Nutzung und Planung der Unternehmensflexibilität. Dissertation RWTH Aachen

[Scha-93]   Scharf D (1993) Grundzüge des betrieblichen Rechnungswesens. Gabler Verlag, Wiesbaden

[Schm-95]   Schmigalla H (1995) Fabrikplanung – Begriffe und Zusammenhänge. REFA München, Hanser Verlag, München

[Schm-96]   Schiemenz B (1996) Komplexität von Produktionssystemen. In: Kern W, Schröder H-H, Weber J (Hrsg.) Handwörterbuch der Produktionswirtschaft, 2. Auflage, Stuttgart

[Schn-01]   Schneeweiß C (2001) Einführung in die Produktionswirtschaft, vol 8, Auflage. Springer Verlag, Berlin u. a

[Scho-88]   Schott G (1988) Kennzahlen: Instrument der Unternehmensführung, vol 5, Auflage. Forkel-Verlag, Wiesbaden

[Schü-94]   Schüpbach H (1994) Prozessregulation in rechnerunterstützten Fertigungssystemen. Vieweg Verlag, Wiesbaden

[Seic-01]   Seicht G (2001) Moderne Kosten- und Leistungsrechnung, vol 11, Auflage. Linde Verlag, Wien

[SeSe-90]   Sethi AK, Sethi SP (1990) Flexibility in manufacturing: a survey. Int J Flex Manuf Syst 2(4):289–328

[Sest-03]   Sesterhenn M (2003) Bewertungssystematik zur Gestaltung struktur- und betriebsvariabler Produktionssysteme. Dissertation RWTH Aachen

[Seyb-81]   Seybert AF (1981) Estimation of damping from response spectra. J Sound Vib 75(2):199–206

[SGWK-04] Schuh G, Gulden A, Wemhöner N, Kampker A (2004) Bewertung der Flexibilität von Produktionssystemen. In: wt Werkstattstechnik online, Jahrgang 94, Heft 6, Springer-VDI-Verlag, Düsseldorf

[ShMo-98]   Shewchuk JP, Moodie CL (1998) Definition and classification of manufacturing flexibility types and measures. Int J Flex Manuf Syst 10(4):325–349 (25)

[SoPa-87]   Son YK, Park CS (1987) Economics measure of productivity, quality and flexibility in advanced manufacturing systems. J Manuf Syst 6(3):194, ff

[Spur-97]   Spur G (1997) Optionen zukünftiger industrieller Produktionssysteme. Interdisziplinäre Arbeitsgruppen – Forschungsberichte, Band 4, Akademischer Verlag, Berlin

[Stau-05]   Staud JL (2005) Datenmodellierung und Datenbankentwurf- Ein Vergleich aktueller Methoden. Springer Verlag, Berlin

[Stau-85]   Staudt E (1985) Kennzahlen und Kennzahlensysteme – Grundlagen zur Entwicklung und Anwendung. Schmidt Verlag, Berlin

[SWF-05]   Schuh G, Wernhörner N, Friedrich C (2005) Lifecycle oriented evaluation of automotive body shop flexibility. In: Zäh, M.F. u.a. (Hrsg.) CARV 2005, München, S.433–S.439

[Temp-93]   Tempelmeier H (1993) Flexible Fertigungssysteme: Entscheidungsunterstützung für Konfiguration und Betrieb. Springer Verlag, Berlin, Heidelberg u. a

[Teum-04]   Teumer H (2004) Wirtschaftliche Produktion am Standort Deutschland durch wandlungsfähige Organisationsformen. In: 5. Deutsche Fachkonferenz Fabrikplanung, Tagungsband, 31 März–1 April 2004, Stuttgart

[Trös-05]   Tröster F (2005) Steuerungs- und Regelungstechnik für Ingenieure, vol 2, Auflage. Oldenbourg Wissenschaftsverlag, München, Wien

[UlHi-76]   Ulrich P, Hill W (1976) Wissenschaftstheoretische Grundlagen der Betriebswirtschaftlehre. Teil 1/2, WiSt Wirtschaftswissenschaftliches Studium, Heft 7/8, Juli/August 1976.

[Ulri-81]   Ulrich H (1981) Die Betriebswirtschaftslehre als anwendungsorientierte Sozialwissenschaft. In: Geist N, Köhler R (Hrsg.) Die Führung des Betriebes. Pöschel Verlag, Stuttgart

[Upto-95]   Upton DM (1995) Flexibility as process mobility: the management of plant capabilities for quick response manufacturing. J Oper Manage 12:205–224

[Upto-97]   Upton DM (1997) Process range in manufacturing. An empirical study of flexibility. Manage Sci 43(8):S.1079–S.1092
[VaSi-04]   Vahrenkamp R, Siepermann C (2004) Produktionsmanagement, vol 5, Auflage. Oldenbourg Wissenschaftsverlag, München
[VDI-95]   Verein Deutscher Ingenieure(1995) Wertanalyse. Idee, Methode, System. 5. Auflage, VDI-Zentrum Wertanalyse, VDI Verlag, Düsseldorf
[Volb-81]   Volberg K (1981) Zur Problematik der Flexibilität manueller Arbeit. Dissertation Universität Düsseldorf
[Warn-93]   Warnecke HJ (1993) Die Fraktale Fabrik – Revolution in der Unternehmenskultur, vol 2, Auflage. Springer-Verlag, Berlin
[WaSt-04]   Waldmann KH, Stocker UM (2004) Stochastische modelle. Springer Verlag, Berlin
[Webe-95]   Weber J (1995) Logistik-controlling. Schäffer-Poeschel Verlag, Stuttgart
[Wegn-97]   Wegner U (1997) Organisation der logistik. Erich Schmidt Verlag, Berlin
[West-04]   Westkämper E (2004) Hochlauf von Fabriken und Produktionssystemen. Wirtschaftliche Produktion am Standort Deutschland durch wandlungsfähige Organisationsformen. In: 5. Deutsche Fachkonferenz Fabrikplanung, Tagungsband, 31 März– 1 April 2004, Stuttgart
[West-06]   Westkämper E (2006) Einführung in die organisation der produktion. Strategien der produktion. Springer Verlag, Berlin, Heidelberg
[WHG-02]   Wiendahl H-P, Hernández R, Grienitz V (2002) Planung wandlungsfähiger Fabriken. ZWF 97 (2002) Heft 1–2
[WHK-06]   Wurst K-H, Heisel U, Kircher C (2006) (Re)konfigurierbare Werkzeugmaschinen – notwendige Grundlage für eine flexible Produktion. In: wt Werkstatttechnik, Jahrgang 96, Heft 5, Springer-VDI-Verlag, Düsseldorf
[Wien-02]   Wiendahl H-P (2002) Die Zukunft prognostizieren mit Szenarien. In: New Management, Band 71, Heft 5
[Wild-05]   Horst Wildemann H (2005) Betreibermodelle: Ein Beitrag zur Steigerung der Flexibilität von Unternehmen? In: Kaluza B, Blecker T (Hrsg.) Erfolgsfaktor Flexibilität. Erich Schmidt Verlag, Berlin
[Wild-87]   Wildemann H (1987) Investitionsplanung und Wirtschaftlichkeitsrechnung für flexible Fertigungssysteme (FFS). Schäffer Verlag, Stuttgart
[Wild-98a]   Wildemann H (1998) Die modulare Fabrik: Kundennahe Produktion durch Fertigungssegmentierung, vol 5, überarb. und erg. Auflage. TCW-Transfer-Centrum, München
[Wild-98b]   Wildemann H (1998) Das agile Unternehmen. Kostenführerschaft und Service, Tagungsband Münchner Management Kolloquium. TCW, München
[WJR-92]   Womack JP, Jones DT, Roos D (1992) Die zweite Revolution in der Autoindustrie: Konsequenzen aus der weltweiten Studie aus dem Massachusetts Institute of Technology. 6. Auflage, 1992, Frankfurt am Main.
[WNKB-05]   Wiendahl HP, Nofen D, Klußmann JH, Breitenbach F (2005) Planung modularer Fabriken – Vorgehen und Beispiele aus der Praxis. Hanser Verlag, Wien, München
[Wolf-90]   Wolf J (1990) Investitionsplanung zur Flexibilisierung der Produktion. Dissertation Technische Hochschule Darmstadt
[WWL-05]   Wahab MIM, Wu D, Lee C-G (2005) A generic approach to measuring the machine flexibility of a manufacturing system. Department of Mechanical and Industrial Engineering, University of Toronto
[ZaDi-94]   Zahn E, Dillerup R (1994) Fabrikstrategien und – strukturen im Wandel. In: Zülch, G. (Hrsg.) Arbeitspapier in Vereinfachen und Verkleinern – Die neuen Strategien in der Produktion, Stuttgart
[Zahn-94]   Zahn E (1994) Produktion als Wettbewerbsfaktor. In: Corsten H (Hrsg.) Handbuch Produktionsmanagement: Strategie – Führung – Technologie – Schnittstellen. Gabler Verlag, Wiesbaden

[ZäMü-07]   Zäh MF, Müller N (2007) On planning and evaluating capacity flexibilities in uncertain markets. 2. International Conference on Changeable, Agile, Reconfigurable and Virtual Production (CARV 2007) Toronto/Canada

[Zäpf-07]   Zäpfel G (2007) Taktisches produktionsmanagement, vol 2, Auflage. Oldenbourg Wissenschaftsverlagsverlag, München

[ZBM-06]   Zäh MF, von Bredow M, Möller N (2006) Methoden zur Bewertung von Flexibilität in der Produktion. In: Industrie Management 22 (2006) 4, GITO-Verlag, Berlin

[ZWL-99]   Zölch M, Weber W, Leder E (1999) Praxis und Gestaltung kooperativer Arbeit. vdf Hochschulverlag AG, Zürich

Index